Sequences and Series

Sequences and Series

Dr. Diamantis Ioannis

Central West Publishing

This edition has been published by Central West Publishing, Australia

For more information about the books published by Central West Publishing, please visit https://centralwestpublishing.com

Disclaimer
Every effort has been made by the publisher and author while preparing this book, however, no warranties are made regarding the accuracy and completeness of the content. The publisher and author disclaim without any limitation all warranties as well as any implied warranties about sales, along with fitness of the content for a particular purpose. Citation of any website and other information sources does not mean any endorsement from the publisher and authors. For ascertaining the suitability of the contents contained herein for a particular lab or commercial use, consultation with the subject expert is needed. In addition, while using the information and methods contained herein, the practitioners and researchers need to be mindful for their own safety, along with the safety of others, including the professional parties and premises for whom they have professional responsibility. To the fullest extent of law, the publisher and author are not liable in all circumstances (special, incidental, and consequential) for any injury and/or damage to persons and property, along with any potential loss of profit and other commercial damages due to the use of any methods, products, guidelines, procedures contained in the material herein.

NATIONAL LIBRARY OF AUSTRALIA

A catalogue record for this book is available from the National Library of Australia

ISBN (print): 978-1-925823-19-6

To my wife Marianna

&

my brother Nektarios

Preface

This book grew out of lectures the author gave in China Agricultural University during Spring 2018 semester. The aim of this book is to serve as a reference for an introduction to undergraduate courses such as "Calculus II" and "Introductory Real Analysis". These courses serve as a foundation for future mathematics as well as studies in other disciplines, particularly in economics and statistics, and with this textbook we provide the background required for insight into more advanced courses both in pure and applied mathematics. Junior & senior mathematics majors, as well as educators or even people with an interest in mathematics can benefit from these lecture notes. We assume that the reader is familiar with very few basic concepts of "Calculus I", such as continuity & differentiability of real functions and some basic integration techniques.

Chapter 1 is concerned with the real number system and it is the author's intention to briefly present the axioms of this number system and emphasize on the most interesting one, completeness. Chapter 2 treats sequences and our purpose is to focus on the concept of convergence. The Bolzano-Weierstrass theorem for sequences and their subsequences, as well as the Cauchy sequences and their properties are discussed in this Chapter, which is concluded by relating continuity of real functions to convergent sequences. Chapter 3 deals with the study of infinite series with a focus on tests for convergence. The Cauchy product of two series is also presented at the end of this chapter. Chapter 4 deals with point-wise and uniform convergence, before Chapter 5, in which our main focus is Power Series and its applications. The concept of approximation of a function with polynomials is studied and applications of this beautiful theory are presented before we proceed with the final chapter of this book, the Infinite Products. This final chapter comprises an introduction to the representation of functions by infinite processes. Moreover, some important exercises can be found at the end of each paragraph and at the end of each chapter and students or even interested readers are encouraged to try to solve them. They range from routine to very difficult. Corrections–mathematical and typographical–are welcome and will be incorporated when received.

It is the author's belief that everyone can benefit from the mathematical maturity that can be gained from these notes. With this textbook we aim for the achievement of the following learning goals:

- study in detail mathematical definitions and statements;
- develop mathematical maturity and improve written communication skills with a particular focus on the clarity, accuracy and completeness of analytical and logical arguments.

These are critical steps to become a mathematician or scientific scholar ready for graduate studies.

Beijing 2018,
Ioannis Diamantis

Acknowledgments

I express my gratitude to everyone who directly or indirectly helped me finish this book and especially my wife Marianna Press for her immerse support and for being a constant source of encouragement. I would also like to express my gratitude to my parents, Athanasios and Freideriki and to my brothers, Nektarios and Emmanouil for the unconditional love and care they provided me through my entire life.

I would like to place on record my sense of gratitude to my good friends Nikolaos Kovatsos and Elena Karli for their constant support and help.

I must acknowledge all the people at China Agricultural University and in particular the people at International College Beijing, and especially Yanru Chen for giving me the freedom to create this new course and for always providing all necessary support.

Finally, I am indebted to my students who showed interest in this beautiful mathematical theory and did their best to succeed in this demanding in many aspects course. This is one of the most satisfying experience a professor may have.

The mathematical sciences particularly exhibit order, symmetry and limitation; and these are the greatest forms of the beautiful.

Aristotle

Contents

Chapter 1

The Real Numbers

In this initial chapter we present the background needed for the study of sequences and series. In addition to introducing basic concepts, this chapter also aims in presenting some proofs with a focus on the clarity, accuracy and completeness of analytical and logical arguments, something necessary for the study of any mathematical theory.

In §1.3 we establish the notation and provide definitions that will be used throughout this textbook. In particular, we study sets and subsets of $\mathbb{R}$ and define the concept of *supremum* and *infimum* of a set. It is worth mentioning that we will naively regard sets as *well-defined* collections of elements and we won't present a list of axioms for set theory. This definition suffices for the purpose of this textbook and the reader is referred to other textbooks on "Set Theory" for an axiomatic approach. The most important result in this section is that the set of rational numbers is *dense* in $\mathbb{R}$. Finally, in §1.4 we present a very important method of proof, Mathematical Induction, and we prove some particular types of statements that we will use throughout the book.

1.1 Introduction

We will regard a *set* as a collection of elements. More precisely, sets are *well-defined* collection of objects, where "well-defined" means that there is a definite method to determine whether an object belongs to the set or not (see Exercises 1.3.13 - 9). Note that the order or repetition of the elements are immaterial.

Notation of the form $\{x|x \text{ satisfies the property } Q\}$ is used for specifying the set of all elements x that satisfy property Q (**rule method**). Finite sets are also specified by explicitly listing their members between braces (**tabulation method**). The number of elements in a set A, denoted $|A|$, is called the *cardinality* of the set. A set with no elements (i.e., cardinality equals 0) is called the *empty set* and is denoted by $\emptyset$, that is $\emptyset := \{\ \}$. The empty set plays the role of the zero element in addition of real numbers in the sense that $A\bigcup\emptyset = A$, and $A\bigcap\emptyset = \emptyset$, $\forall A$, just like $a + 0 = 0$ and $a \cdot 0 = a, \forall a \in \mathbb{R}$.

A subset B of A, denoted as $B \subset A$, is a collection of (some) elements that belong to A. That is:

$$B \subset A \text{ if} - \text{f } \forall x \in B \Rightarrow x \in A.$$

Definition 1.1.1. We define operations on sets as follows:

i. The *union* of A and B, denoted $A \cup B$, is the set:

$$\{x|x \text{ is either in A or in B}\}.$$

ii. The *intersection* of A and B, denoted $A \cap B$, is the set:

$$\{x|x \text{ is both in A and in B}\}.$$

iii. A and B are said to be *disjoint* if they have no element in common, that is, if $A \cap B = \emptyset$.

iv. The *difference* between A and B, denoted $A - B$, is the set:

$$\{x|x \text{ is in A but not in B}\}.$$

v. If B is a subset of A, then $A - B$ is said to be the *complementary* of B with respect to A, denoted $\overline{B}$ or B^c.

1.2 The Real Numbers $\mathbb{R}$

Let $a, b \in \mathbb{R}$. Then the following properties can be used to manipulate expressions (as well as real numbers):

$$\begin{array}{lcll}
a+b & = & b+a & \text{Commutative Property of addition} \\
a+(b+c) & = & (a+b)+c & \text{Associative Property of addition} \\
a+0 & = & a & \text{Additive identity Property} \\
a+(-a) & = & 0 & \text{Additive inverse Property} \\
a\cdot b & = & b\cdot a & \text{Commutative Property of multi.} \\
a\cdot(b\cdot c) & = & (a\cdot b)\cdot c & \text{Associative Property of multi.} \\
a\cdot 1 & = & a & \text{Multiplicative identity Property} \\
a\cdot \frac{1}{a} & = & 1 & \text{Multiplicative inverse Prop., } \forall a \neq 0 \\
a\cdot(b+c) & = & a\cdot b + a\cdot c & \text{Distributive Property}
\end{array}$$

Exercise 1.2.1. When there is one and only one element with a given property, we say that the element is *unique*. Show that *zero*, the identity element for addition, is unique.

We are now in position to present a set of *axioms* that define and characterize the set of real numbers:

Definition 1.2.2. The set of *real numbers* consist of elements that satisfy the following properties:

- The *closure* axioms: $a + b \in \mathbb{R}$ and $a \cdot b \in \mathbb{R}, \forall a, b \in \mathbb{R}$.
- Properties of addition.
- Properties of multiplication.

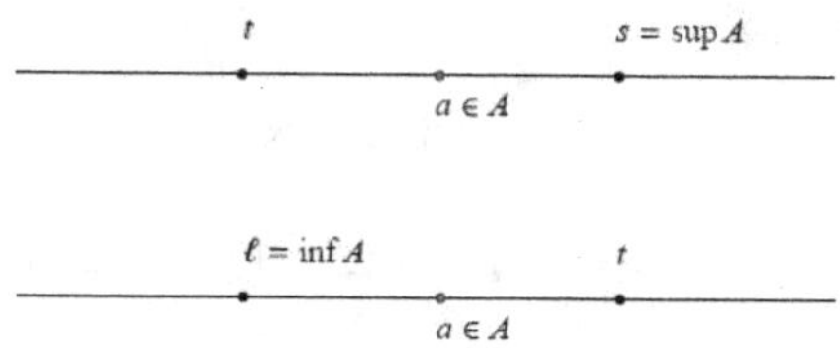

- The *trichotomy* axiom: $\forall a, b \in \mathbb{R}$, one and only one of the following statements is true: $a > b$, or $a = b$, or $a < b$.
- *Transitive* axiom: If $a > b$ and $b > c$, then $a > c$.
- *Addition* axiom: If $a > b$, then $a + c > b + c, \forall c \in \mathbb{R}$.
- *Multiplication* axiom: If $a > b$, then $a \cdot c > b \cdot c, \forall c \in \mathbb{R}^+$.
- The *completeness property*.

In the next subsection we present the completeness property in order to have a precise definition of the set of real numbers.

1.3 Bounded subsets of $\mathbb{R}$: Supremum and Infimum

Definition 1.3.1. Let $A \subset \mathbb{R}$ such that $A \neq \emptyset$.

i. A is *bounded above* if $\exists\, u \;:\; x \leq u,\ \forall x \in A$

ii. A is *bounded below* if $\exists\, v \;:\; v \leq x,\ \forall x \in A$

iii. A is *bounded* if $\exists\, u, v \;:\; v \leq x \leq u,\ \forall x \in A$

Remark 1.3.2. If A is bounded above by u and $u < m$, then A is bounded above by m. That is, if a set is bounded above, then it has infinite many upper bounds. Consequently, if a set is bounded below, then it has infinite many lower bounds.

Definition 1.3.3. Let $A \subset \mathbb{R}$ such that $A \neq \emptyset$ and A bounded. Then:

i. The least upper bound of A is called *supremum* of A, denoted as $\sup A$.

ii. The greatest lower bound of A is called *infimum* of A, denoted as $\inf A$.

It is worth mentioning that a bounded set doesn't necessarily contain a minimum or a maximum element. By Definition 1.3.3 though, a bounded set always has a supremum and an infimum. The following example sheds more light to this fact.

Example 1.3.4. i. If $A = [a, b]$, then $\min\{A\} = a = \inf A$, and $\max\{A\} = b = \sup A$.

ii. For $A = (a, b)$, then $\nexists\ \min\{A\}, a = \inf A, \nexists\ \max\{A\}, b = \sup A$.

iii. If $A = (a, +\infty)$, then $\nexists\ \min\{A\}, a = \inf A, \nexists\ \max\{A\}$, but

$$\sup A = \infty.$$

iv. For $A = \left\{\frac{1}{n},\ n \in \mathbb{N}\right\}$, we have: $\nexists\ \min\{A\}, 0 = \inf A$, and

$$\max\{A\} = 1 = \sup A.$$

Example 1.3.5. i. We say that tow sets A and B are *equal* if and only if $A \subseteq B$ and $B \subseteq A$. Note that if

$$\inf A = \inf B \text{ and } \sup A = \sup B,$$

then we can't conclude that $A = B$. See Example 1.3.4.

ii. $\sup A \leq \inf B \Leftrightarrow x \leq y,\ \forall\, x \in A\ \&\ y \in B.$

Remark 1.3.6. Let $A \neq \emptyset$, a bounded subset of $\mathbb{R}$. Then $\exists\ \sup A, \inf A$ in $\mathbb{R}$ and:

$$\forall \epsilon > 0\, \exists\, x \in A : \sup A - \epsilon < x \leq \sup A \tag{1.1}$$

$$\forall \epsilon > 0\, \exists\, x \in A : \inf A \leq x < \inf A + \epsilon \tag{1.2}$$

We will refer to (1.1) as the *property of the supremum* and to (1.2) as the *property of the infimum*.

The Completeness Property of $\mathbb{R}$: *Every nonempty set of real numbers that has an upper bound also has a supremum.*

Theorem 1.3.7. *The set of natural numbers* $\mathbb{N} := \{1, 2, 3, \ldots, n, \ldots\}$ *is not bounded above, i.e.* $\sup \mathbb{N} = +\infty$.

Proof. Let $\mathbb{N}$ be bounded above. Then, $\exists\ \sup \mathbb{N} = n \in \mathbb{N}$. Since now $n - 1$ is not an upper bound of $\mathbb{N}$, from the property that characterizes the supremum, $\exists\ n_1 \in \mathbb{N}\ :\ \sup \mathbb{N} - 1 < n_1 \Rightarrow \sup \mathbb{N} < n_1 + 1$, which leads to a contradiction, since $n_1 + 1 \in \mathbb{N}$. □

Theorem 1.3.8. *Archimedean Property: If* $l > 0,\ \exists\, n \in \mathbb{N}\ :\ 0 < \frac{1}{n} < l.$

Proof. We prove the Archimedean property by contradiction. Assume that Thm. 1.3.8 is not true. Then, if there is no such n then $1/n < l$ for *all* integers n. That is, there would be an upper bound for the integers. This contradicts Theorem 1.3.7. □

We now proceed with defining the *integer part of a real number* and for that we first present the following proposition and we omit the proof.

Proposition 1.3.9. $\forall x \in \mathbb{R},\ \exists\,!\, k \in \mathbb{Z}\ :\ k \leq x < k + 1.$

Definition 1.3.10. Let $k \in \mathbb{Z}\ :\ k \leq x < k + 1$. The integer k is called the *integer part* of x and is denoted as $\lfloor x \rfloor$.

We are now in position to prove the *density* of rational numbers in $\mathbb{R}$.

Theorem 1.3.11. *The set of rational numbers* $\mathbb{Q} := \left\{\frac{p}{q}, \mid q \neq 0\right\}$ *is* <u>*dense*</u> *in* $\mathbb{R}$. *That is:* $\forall\, a < b\ \exists\, r \in \mathbb{Q}\ :\ a < r < b.$

Proof. Since $a < b$, from the Archimedean property we have that

$$\exists\, n \in \mathbb{N}\ :\ \frac{1}{n} < b - a \Rightarrow 1 + na < nb \Rightarrow$$
$$na < \lfloor na \rfloor + 1 < na + 1 < nb \Rightarrow$$
$$a < \underbrace{\frac{\lfloor na \rfloor + 1}{n}}_{\mathbb{Q}} < b$$

□

Example 1.3.12.

1. Let $A, B \subset \mathbb{R}$ bounded such that $\inf A < \inf B$. We show that $\exists\, a \in A\ :\ a = \inf B$. Let $\epsilon = \inf B - \inf A > 0$. Then, $\exists\, a \in A\ :$
$$a < \inf A + \epsilon = \inf A + \inf B - \inf A = \inf B.$$
2. Let $A, B \subset \mathbb{R}$ bounded such that $\sup A = \inf B$. We show that $\forall\, \epsilon > 0,\ \exists\, a \in A\ \&\ b \in B\ :\ b - a < \epsilon$. Let $\epsilon > 0$. Then,
$$\exists\, a \in A\ \&\ b \in B\ :\ a > \sup A - \frac{\epsilon}{2}\ \&\ b < \inf B + \frac{\epsilon}{2} \Rightarrow$$
$$b - a < \inf B + \frac{\epsilon}{2} - \left(\sup A - \frac{\epsilon}{2}\right) = \epsilon$$

Exercises 1.3.13. Sets, Supremum & Infimum

1. Let $A, B \subset \mathbb{R}$ bounded such that $\inf A < \sup B$. Prove that
$$\exists\, a \in A\ \&\ b \in B\ :\ a < b.$$
2. Let $A, B \subset \mathbb{R}$ bounded such that $\sup A < \sup B$. Prove that
$$\exists\, b \in B\ :\ b = \sup A.$$
Would it be true if instead of $\sup A < \sup B$ we had $\sup A \leq \sup B$?
3. Let $A, B \subset \mathbb{R}$ bounded such that $\sup A = \inf B$. Prove that
$$\forall\, \epsilon\ \exists\, a \in A\ \&\ b \in B\ :\ b - a < \epsilon.$$
4. Let $A, B \subset \mathbb{R}$ bounded such that $\inf A < \inf B$. Prove that
$$\exists\, a \in A\ :\ a \inf B.$$
5. Prove that every set of real numbers that is bounded below, it has ab infimum.
6. Prove that the supremum of a bounded above set is unique.
7. Let $A \subset \mathbb{R}$ that is bounded above and define
$$b + A := \{b + a, a \in A\},\ \text{where} b \in \mathbb{R}.$$
Prove that $\sup(b + A) = b + \sup A$.
8. Prove that $A - (A \cap B) = A - B$ and $(A \cup B) - B = A - B$.
9. (Russell's paradox) Explain why the "set" $A = \{a\ :\ a \notin A\}\}$ does not fit the definition of sets presented in the Introduction.

1.4 Mathematical Induction

The Principle of Mathematical Induction:

Let S_n be a statement involving the positive integer n. If:

1. S_1 is true, and
2. the truth of the statement S_k implies the truth of the statement S_{k+1}, for every positive integer k,

then the statement S_n is true for all positive integers n.

Remark 1.4.1. To prove that S_n is true for all positive integers n requires two steps:

Base of induction: Show that S_1 is true.

Induction hypothesis: Assume that S_k is true and prove S_{k+1} is also true.

Example 1.4.2. We prove that $S_n = 1+2+3+\ldots+n := \sum_{i=1}^{n} i = \frac{n(n+1)}{2}$, for all $n \in \mathbb{N}$. This was proved by *Gauss* when he was a high school student and we will refer to this relation as the **Gauss relation**.

Base of induction: For $n = 1$ we have that

$$S_1 = 1 := \sum_{i=1}^{i} i = \frac{1(1+1)}{2} = 1.$$

So the relation holds for $n = 1$.

Induction hypothesis: Assume that the relation holds for $n = k \in \mathbb{N}\backslash\{1\}$, that is $S_k = 1+2+3+\ldots+k := \sum_{i=1}^{k} i = \frac{k(k+1)}{2}$.

We will prove that it is true for $k+1$. Indeed we have that:

$$\begin{aligned} S_{k+1} &= 1+2+3+\ldots+k+k+1 := \sum_{i=1}^{k+1} i = \\ &= \sum_{i=1}^{k} i + (k+1) \overset{ind.hyp.}{=} \\ &= \frac{k(k+1)}{2} + k+1 = \frac{(k+1)(k+2)}{2} \end{aligned}$$

Thus, the relation holds for all $n \in \mathbb{N}$.

Example 1.4.3. The following inequality holds for all $n \in \mathbb{N}$ such that $n \geq 2$ and for all $-1 < a \neq 0$:

$$(1+a)^n > 1 + na \tag{1.3}$$

This inequality is known as the **Bernoulli inequality**.

Base of induction: For $n = 2$ we have that $(1+a)^2 > 1+2a \Leftrightarrow 1+2a+a^2 > 1+2a \Leftrightarrow a^2 > 0$, which holds.

Induction hypothesis: Assume that the relation holds for $n = k \in \mathbb{N}$ where $k > 2$. That is $(1+a)^k > 1+ka$.

We prove that it is true for $k+1$. Indeed we have:

$$(1+a)^k > 1+ka \Leftrightarrow (1+a)(1+a)^k > (1+a)\cdot(1+ka) \Leftrightarrow$$

$$(1+a)^{k+1} > 1+ka+a+ka^2 \Leftrightarrow$$

$$(1+a)^{k+1} > 1+(k+1)a+ka^2 \overset{ka^2>0}{\Leftrightarrow}$$

$$(1+a)^{k+1} > 1+(k+1)a$$

Thus, the inequality holds for all $n \in \mathbb{N}$ such that $n \geq 2$.

Exercises 1.4.4. Use **mathematical induction** to prove the following:

1. $\sum_{i=1}^{n} i^2 := 1^2+2^2+3^2+\ldots+n^2 = \frac{n(n+1)(2n+1)}{6}$.

2. $\sum_{i=1}^{n} i^3 := 1^3+2^3+3^3+\ldots+n^3 = \frac{n^2(n+1)^2}{4}$.

3. $3^n > n^2$ for $n \in \mathbb{N}$.

4. n^3+2n is divisible by 3, for any positive integer n.

5. $n! > 2^n$, for $n \in \mathbb{N}$ such that $n > 4$.

6. Prove that for all $n \in \mathbb{N}$, $\sum_{k=1}^{n} 2k-1 = n^2$.

Review Exercises

1. Prove that $A - (A \cap B) = A - B$ and $(A \cup B) - B = A - B$.

2. Let A and B be subsets of a set X. Show that $(A \cup B)^c = A^c \cap B^c$. What is $(A \cap B)^c$?

3. Let $A = \{x \in \mathbb{Q} : x < \sqrt{2}\}$. Does A attain a supremum? Explain.

4. Find the least upper bound of the set $\{\frac{1}{2}, \frac{2}{3}, \frac{3}{4}, \ldots\}$ and prove your answer.

5. We define $n!$ inductively for $n = 0, 1, 2, \ldots$ by $0! = 1$ and $n! = n \times (n-1)!$. For $n, k = 0, 1, 2, \ldots$, define $\binom{n}{k}$ by
$$\binom{n}{k} = \frac{n!}{k!(n-k)!}.$$
Prove that
$$\binom{n+1}{k} = \binom{n}{k-1} + \binom{n}{k}.$$

6. Prove that for all $x \in \mathbb{R}$, $0 \leq |x| - x \leq 2|x|$.

7. Let $x \in \mathbb{R}$ and $n \in \mathbb{N}$. Prove that $\exists\, m \in \mathbb{Z} : \left|x - \frac{m}{n}\right| < \frac{1}{n}$.

8. Let A be a set, and define the **power set** of A, denoted as$\mathcal{P}(A)$, to be the set of all subsets of A, that is:
$$\mathcal{P}(A) = \{B \mid B \subseteq A\}.$$
Prove that the number of subsets of the set $[[n]] = \{1, 2, \ldots, n\}$ is 2^n.

9. A subset A of $\mathbb{Q}$ is said to be a *cut* (or Dedekind cut) if it satisfies the following:

   (a) $A \neq \emptyset$ and $A \neq \mathbb{Q}$

   (b) If $r \in A$ and $s \in \mathbb{Q}$ satisfies $s < r$, then $s \in A$

   (c) If $r \in A$ then there is some $s \in \mathbb{Q}$ with $s > r$, $s \in A$.

   Prove that $\{x \in \mathbb{Q} \mid x \leq 0\}$ is not a Dedekind cut, while
$$\{x \in \mathbb{Q} \mid x < 0\} \cup \{x \in \mathbb{Q} \mid x^2 < 2\}$$
is a Dedekind cut.

10. An *abelian group* is a set A with an operation $\star : A \times A \to A$ that satisfies the following properties:

    - (Commutativity) $a \star b = b \star a$ for every $a, b \in A$.
    - (Associativity) $(a \star b) \star c = a \star (b \star c)$ for every $a, b, c \in A$.
    - (Identity) There exists an element $e \in A$ so that $a \star e = a$ for every $a \in A$.
    - (Inverses) For each $a \in A$, there is an element $a' \in A$ so that $a \star a' = e$.

For example, the integers $\mathbb{Z}$ with addition as the operation is an abelian group. On the other hand, $\mathbb{Z}$ under multiplication and the natural numbers $\{0, 1, 2, \ldots\}$ under addition are not abelian groups because not every element has an inverse.

Let X be a set. For subsets $A \subset X$ and $B \subset X$ the *symmetric difference* is defined as

$$A \Delta B = (A - B) \cup (B - A).$$

We can view this as an operation $\Delta : P(X) \times P(X) \to P(X)$. Which of the abelian group axioms does this operation satisfy?

11. For each of the following subsets of $\mathbb{R}$, write down

$$\inf = \qquad \min = \qquad \sup = \qquad \max =$$

and fill in the correct values.

- $(-1, 2) \cup [3, 4]$
- $\{\frac{3^n}{n} : n \in \mathbb{N}\}$

Chapter 2

Sequences

In this chapter we study the convergence of *sequences*. We will build a beautiful theory with the use of which, many results of "Calculus I" can be proved (see § 2.11 for the essence of continuity via sequences). An introduction to the theory of sequences is usually presented in courses such as "Calculus II", but the way we will treat this subject is more suitable for courses such as "Real Analysis".

We first introduce the essence of sequences and discuss bounded sequences and monotonic sequences. The definition and meaning of convergence of a sequence of real numbers is then presented together with some useful results. As in the case of most mathematical concepts, the formal definition of convergence is sometimes hard to use and one needs to find other tools to study this concept. The rest of this chapter deals with the presentation of some necessary tools and some deeper results concerning the convergence of sequences. Cauchy sequences are finally treated in § 2.10.

2.1 Introduction

A sequence can be regarded as a list of numbers written in a definite order

$$\{a_1, a_2, \ldots, a_n, \ldots\},$$

where a_1 is called the first term of the sequence, a_2 the second term of the sequence and a_n is called the nth term or the *general term* of the sequence. We will deal exclusively with *infinite sequences* and so each term a_n will have a successor a_{n+1}. This leads to the following definition:

Definition 2.1.1. An *infinite sequence* (or just sequence) a_n can be defined as a function whose domain is the set of natural numbers $\mathbb{N}$. Sequences whose domains consist only of the first n natural numbers are called *finite sequences*.

Remark 2.1.2. Let us remind the reader the definition of a real function: A *function* f is a rule that assigns to each value x in a set D a unique value denoted $f(x)$. The set D is called the *domain* of the function. The *range* is the set of all values of $f(x)$ produced as x varies over the domain. Note that instead of the standard notation for functions, $f(n)$, we will use the notation $(a_n)_{n\in\mathbb{N}}$ or $\{a_n\}_{n=1}^{\infty}$ for sequences.

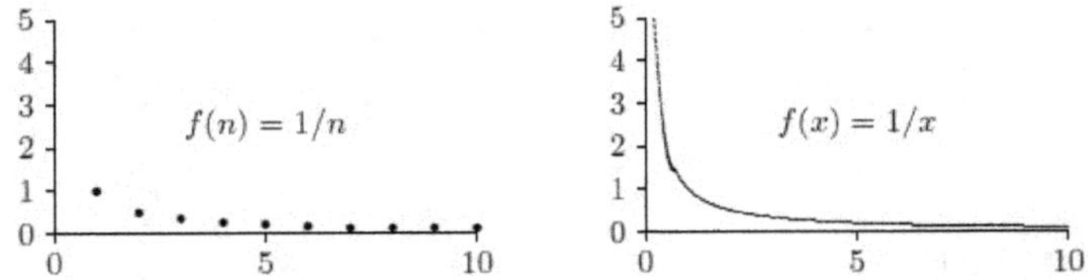

Figure 2.1: Sequences and functions

It is very convenient to have the general term of a sequence, for example: $a_n = \frac{n}{n+1}$, but a sequence may be defined *recursively* (by a recursion formula). Such a definition lists the first term or the first few terms and then describes hot to determine the remaining terms from the given terms. For example, $a_1 = 2$ and $a_{n+1} = a_n^2 - 3$ for all $n \in \mathbb{N}$.

We now define the factorial of a natural number.

Definition 2.1.3. If n is a natural number, the notation $n!$ (read "n factorial") is the product of all natural numbers from n down through 1. That is:

$$n! \ := \ n \cdot (n-1) \cdot \ldots \cdot 2 \cdot 1$$

We also define zero factorial to be equal to 1, $0! := \ 1$.

Example 2.1.4. i. Let $a \in \mathbb{R}$. The sequence $A = \{a, a, \ldots\}$ is called the *constant* sequence.

ii. The *Fibonacci* sequence f_n is given by the induction formula:

$$f_1 := f_2 := 1, \quad f_{n+1} := f_{n-1} + f_n, \ \forall\, n \geq 2.$$

Exercises 2.1.5. Sequences

1. Write the first four terms of the sequence whose n^{th} term, or general term, is given by: $a_n = \frac{n-2}{n^2+1}$.

2. Same for $a_n = \frac{(-1)^n}{n}$.

3. Find the first four terms of the sequence in which $a_1 = 2$ and $a_n = 2a_{n-1} + 4$, for all $n \in \mathbb{N}$.

4. Same for $a_n = \frac{n^2}{n!}$.

5. Rewrite the sequence using bracket notation: $4, 6, 9, 13.5, 23.25, \ldots$

6. Rewrite the sequence using bracket notation: $11, 7, 3, \ldots$

2.2 Arithmetic sequences and series

We now focus on some specific type of sequences called *arithmetic* sequences, where each term is found by adding the same number to the preceding term. More precisely:

Definition 2.2.1. A sequence is called *arithmetic* if $\exists\, d$ called the common difference, such that: $a_{n+1} = a_n + d$ or $a_{n+1} - a_n = d$ for all $n \in \mathbb{N}$.

If for an arithmetic sequence we know the first term and the common difference, then from the recursive formula $a_{n+1} = a_n + d$ (Definition 2.2.1) we can evaluate all terms. More precisely:

$$\begin{array}{lllll} a_1 & & & & \\ a_2 & = & a_1 + d & & \\ a_3 & = & a_2 + d & = & a_1 + 2d \\ \vdots & & & & \vdots \\ a_n & = & a_{n-1} + d & = & a_1 + (n-1)d, \quad \forall\, n \geq 1. \end{array}$$

Thus, we have that n^{th} term of an arithmetic sequence satisfies:

$$a_n = a_1 + (n-1)\, d, \forall\, n \geq 1.$$

Let's consider now three consecutive terms of an arithmetic sequence. Obviously:

$$a_{n+1} - a_n = d, \quad a_n - a_{n-1} = d \quad \Rightarrow a_n = \frac{a_{n+1} + a_{n-1}}{2}.$$

Moreover, if for three real numbers a, b, c we have that $b = \frac{a+c}{2}$, then $a - b = b - c$ and thus, the numbers a, b, c are three consecutive terms of an arithmetic sequence. We have proved the following proposition:

Proposition 2.2.2. *Three real numbers a, b, c are consecutive terms of an arithmetic sequence if-f: $b = \frac{a+c}{2}$.*

Definition 2.2.3. Let a, b, c be consecutive terms of an arithmetic sequence $(a_n)_{n\in\mathbb{N}}$. The number b is called the *arithmetic mean.*

Let now $(a_n)_{n\in\mathbb{N}}$ be an arithmetic sequence. We will now evaluate the sum of the first n terms, S_n. This sum is called the nth *partial sum* of the sequence $(a_n)_{n\in\mathbb{N}}$ and it is the first step towards the study of infinite series.

Proposition 2.2.4. *Let $(a_n)_{n\in\mathbb{N}}$ be an arithmetic sequence. Then:*

$$S_n := a_1 + a_2 + \ldots + a_n = \frac{n}{2} \cdot (a_1 + a_n).$$

Proof. We have that:

$$\begin{aligned} S_n &:= a_1 + a_2 + \ldots + a_n = \\ &= a_1 + (a_1 + d) + (a_1 + 2d) + \ldots (a_1 + (n-1)d) = \\ &= na_1 + d \cdot (1 + 2 + 3 + \ldots + (n-1)) \overset{Gauss}{=} \\ &= na_1 + \tfrac{n(n-1)}{2} \cdot d = \tfrac{n}{2} \cdot (a_1 + a_n) \end{aligned}$$

□

Exercises 2.2.5. Arithmetic Sequences

1. Evaluate the following sum $10 + 13 + \ldots + 163$.
2. The fifth term of an arithmetic sequence is 12 and $a_9 = 28$. Evaluate a_1, d and S_{12}.
3. Evaluate the sum of the first 100 odd numbers.
4. Evaluate the sum $1 + \frac{n-1}{n} + \frac{n-3}{n} + \ldots + \frac{1}{n}$.
5. Rewrite the series using sigma notation: $5 + 15 + 25 + \ldots$

2.3 Geometric sequences and series

We now focus on a different type of sequences called *geometric* sequences, where each term is found by multiplying the preceding term by the same number. More precisely:

Definition 2.3.1. A sequence is called *geometric* if $\exists\, r \neq 1$ called the common ration, such that: $a_{n+1} = a_n \cdot r$ or $\frac{a_{n+1}}{a_n} = r$ for all $n \in \mathbb{N}$.

If now for a geometric sequence we know the first term and the common ration, then from the recursive formula $a_{n+1} = a_n \cdot r$ (Definition 2.3.1) we can evaluate all terms. More precisely:

$$\begin{array}{lllll} a_1 & & & & \\ a_2 & = & a_1 \cdot r & & \\ a_3 & = & a_2 \cdot r & = & a_1 \cdot r^2 \\ \vdots & & & & \vdots \\ a_n & = & a_{n-1} \cdot r & = & a_1 \cdot r^{n-1}, \quad \forall\, n \geq 1. \end{array}$$

Thus, we have that n^{th} term of a geometric sequence satisfies:

$$a_n = a_1 \cdot r^{n-1}, \forall\, n \geq 1.$$

Let's consider now three consecutive terms of a geometric sequence. Obviously:

$$\frac{a_{n+1}}{a_n} = r, \qquad \frac{a_n}{a_{n-1}} = r \quad \Rightarrow \; a_n^2 = a_{n+1} \cdot a_{n-1}.$$

Moreover, if for three real numbers a, b, c we have that $b^2 = a \cdot c$, then $b/a = c/b$ and thus, the numbers a, b, c are three consecutive terms of a geometric sequence. We have proved the following proposition:

Proposition 2.3.2. *Three real numbers a, b, c are consecutive terms of a geometric sequence if-f: $b^2 = a \cdot c$.*

Definition 2.3.3. Let $a, b, c \in \mathbb{R}$ be consecutive terms of a geometric sequence $(a_n)_{n\in\mathbb{N}}$. The number b is called the *geometric mean.*

Let now $(a_n)_{n\in\mathbb{N}}$ be a geometric sequence. We will now evaluate the nth partial sum of the sequence $(a_n)_{n\in\mathbb{N}}$, that is, the sum of the first n terms, S_n.

Proposition 2.3.4. *Let $(a_n)_{n\in\mathbb{N}}$ be a geometric sequence. Then:*

$$S_n := a_1 \cdot \frac{1-r^n}{1-r}, \forall r \neq 1.$$

Proof. We have that:

$$\begin{aligned} S_n &:= a_1 + a_2 + \ldots + a_n = \\ &= a_1 + (a_1 \cdot r) + (a_1 \cdot r^2) + \ldots (a_1 \cdot r^{n-1}) = \\ &= a_1 \cdot (1 + r + r^2 + \ldots + r^{n-1}) \overset{\cdot r}{\Rightarrow} r \cdot S_n = \\ &= r \cdot S_n = a_1 \cdot (r + r^2 + \ldots + r^n) \end{aligned}$$

Now: $S_n - r \cdot S_n = a_1 \cdot (1 - r^n) \overset{r\neq 1}{\Rightarrow} S_n = a_1 \cdot \frac{1-r^n}{1-r}$. □

Example 2.3.5. Let $a_n = 3^n$. Obviously a_n is a geometric sequence and $a_1 = 3$ and $r = 3$. The sum of the first ten terms of the sequence a_n is $S_{10} = 3 \cdot \frac{1-3^{10}}{1-3}$.

The next theorem plays an important role in the theory of infinite series. It deals with the *infinite sum* $\sum_{i=1}^{\infty} a_i := \lim_{n\to\infty} S_n := S_\infty$.

Theorem 2.3.6. *Let $(a_n)_{n\in\mathbb{N}}$ be a geometric sequence with $|r| < 1$. Then:*

$$S_\infty = \frac{a_1}{1-r}.$$

Proof. We have that: $S_\infty := \lim_{n\to\infty} S_n := \lim_{n\to\infty} a_1 \cdot \frac{1-r^n}{1-r} = \frac{a_1}{1-r}$, since $\lim_{n\to\infty} r^n = 0$, for $|r| < 1$. □

Remark 2.3.7. We prove that $\lim_{n\to\infty} r^n = 0$, for $|r| < 1$ in the Chapter 5.

Exercises 2.3.8. Geometric Sequences

1. Evaluate the sum $1 + \frac{1}{2} + \frac{1}{4} \ldots + \frac{1}{256}$.
2. The fifth term of a geometric sequence is 32 and $a_3 = 8$. Evaluate a_1, r, S_{12} and S_∞.
3. Let $a_n = 2^n \cdot \frac{1}{3^{n+1}}$. Prove that a_n is a geometric sequence and evaluate a_1, r and S_{25}.
4. Evaluate the following sums:
   i. $\sum_{n=0}^{\infty} \left(\frac{1}{3}\right)^n$
   ii. $\sum_{n=0}^{\infty} \left(\frac{e}{\pi}\right)^n$
   iii. $\sum_{n=0}^{\infty} \frac{2^{n+3}}{5^n+1}$
5. Using geometric series, prove that $0.999\ldots = 1$.

2.4 Bounded sequences

Definition 2.4.1. Let $(a_n)_{\mathbb{N}}$ be a sequence.

i. a_n is said to be *bounded above* if $\exists\, M \in \mathbb{R} : a_n \leq M,\ \forall n \in \mathbb{N}$.

ii. a_n is said to be *bounded below* if $\exists\, m \in \mathbb{R} : m \leq a_n,\ \forall n \in \mathbb{N}$.

iii. a_n is said to be *bounded* if $\exists\, m, M \in \mathbb{R} : m \leq a_n \leq M,\ \forall n \in \mathbb{N}$.

iv. a_n is said to be *absolutely bounded* if $\exists\, \theta > 0 : |a_n| \leq \theta,\ \forall n \in \mathbb{N}$.

An immediate consequence of Definition 2.4.1 is the following proposition:

Proposition 2.4.2. *A sequence $(a_n)_{\mathbb{N}}$ is bounded if-f $(a_n)_{\mathbb{N}}$ is absolutely bounded.*

Proof. $\Rightarrow$ Let $(a_n)_{\mathbb{N}}$ be a bounded sequence. Then, by Definition 2.4.1 we have that

$$\exists\, m, M \in \mathbb{R} : m \leq a_n \leq M,\ \forall n \in \mathbb{N}.$$

Let $\theta = \max\{|m|, |M|\}$. Then: $\theta \geq |M|$ and $\theta \geq |m|$, that is

$$-\theta \leq M \leq \theta \text{ and } -\theta \leq m \leq \theta.$$

We have that:

$$-\theta \leq m \leq a_n \leq M \leq \theta, \forall n \in \mathbb{N} \Rightarrow |a_n| \leq \theta, \forall n \in \mathbb{N}.$$

$\Leftarrow$ Let $(a_n)_{\mathbb{N}}$ be an absolutely bounded sequence. Then, by Definition 2.4.1 we have that

$$\exists\, \theta > 0 : |a_n| \leq \theta,\ \forall n \in \mathbb{N} \rightarrow -\theta \leq a_n \leq \theta, \forall n \in \mathbb{N}.$$

We conclude that a_n is bounded below by $-\theta$ and bounded above by θ, thus, a_n is bounded.

□

As noted in the introduction, if we regard a sequence $(a_n)_{\mathbb{N}}$ as an infinite ordered list of numbers, that is, the range of $(a_n)_{\mathbb{N}}$ is an infinite subset of $\mathbb{R}$, then it makes perfect sense to check if this subset has a maximum, a minimum, a supremum and an infimum. This leads to the following definition:

Definition 2.4.3. Let $(a_n)_{\mathbb{N}}$ be a bounded sequence.

i. The *least upper bound* of the sequence $(a_n)_{\mathbb{N}}$ is denoted by

$$\sup\{a_n\} := \sup\{a_1, a_2, \ldots, a_n, \ldots\}.$$

ii. The *greatest lower bound* of the sequence $(a_n)_{\mathbb{N}}$ is denoted by

$$\inf\{a_n\} := \inf\{a_1, a_2, \ldots, a_n, \ldots\}.$$

Example 2.4.4. i. Let $a_n = \frac{1}{n}$. Obviously, $0 < a_n \leq 1, \forall\, n \in \mathbb{N}$. Thus, a_n is bounded and $\inf a_n = 0, \sup a_n = 1$.

ii. Let $a_n = \frac{3\sin n - 4\cos n}{n^5}$, $n \in \mathbb{N}$. We have that:

$$\begin{aligned} |a_n| &= \tfrac{1}{n^5} \cdot |3\sin n - 4\cos n| \leq \\ &\leq \tfrac{1}{n^5} \cdot (3|\sin n| + 4|\cos n|) \leq \\ &\leq \tfrac{1}{n^5} \cdot (3+4) = \tfrac{7}{n^5} \leq 7. \end{aligned}$$

Thus, $|a_n| \leq 7 \Rightarrow a_n$ is bounded. Note that 7 is an upper bound of a_n and *not* the supremum of a_n.

Exercises 2.4.5. Bounded Sequences

1. Check whether the following sequences are bounded or not. Find the supremum and infimum of each sequence when these exist.

   i. $a_n = \frac{5\sin n + 7}{n^8 + 1}$
   ii. $a_n = \frac{n!}{n^n}$
   iii. $a_n = \frac{n}{3^n}$ (*Hint:* Use the Bernoulli inequality).
   iv. $a_n = \left(1 + \frac{1}{n}\right)^n$ (*Hint:* Use the Bernoulli inequality).
   v. $a_1 = \sqrt{2}$ and $a_{n+1} = \sqrt{2 + a_n}, \forall\, n \in \mathbb{N}$ (*Hint:* Use induction to prove that $0 < a_n < 2$).

2.5 Monotonic sequences

Definition 2.5.1. Let $(a_n)_{\mathbb{N}}$ be a sequence. We say that a_n is called:

i. *strictly increasing*, denoted by $a_n \nearrow$, if $a_n < a_{n+1}, \forall\, n \in \mathbb{N}$.

ii. *strictly decreasing*, denoted by $a_n \searrow$, if $a_n > a_{n+1}, \forall\, n \in \mathbb{N}$.

iii. *strictly monotonic* if it is strictly increasing or strictly decreasing.

Remark 2.5.2. It is worth mentioning that a sequence $(a_n)_{\mathbb{N}}$ is called *increasing*, if $a_n \leq a_{n+1}$, for all $n \in \mathbb{N}$ and *decreasing* if $a_n \geq a_{n+1}, \forall\, n \in \mathbb{N}$. Moreover, a sequence is said to be *constant* if $a_n = a_{n+1}, \forall\, n \in \mathbb{N}$.

Obviously, a strictly increasing sequence $(a_n)_{\mathbb{N}}$ is bounded below from a_1 and a strictly decreasing sequence $(b_n)_{\mathbb{N}}$ is bounded above from b_1.

Methodology: In order to check whether a sequence $(a_n)_{\mathbb{N}}$ is strictly monotonic and also determine its monotonicity, we work as follows:

i. **The Difference Test**: We consider the difference $a_{n+1} - a_n$ and we compare this difference to zero. Then, we have:

   If $a_{n+1} - a_n > 0 \Rightarrow a_n \nearrow$
   If $a_{n+1} - a_n < 0 \Rightarrow a_n \searrow$
   If $a_{n+1} - a_n = 0 \Rightarrow a_n$ is constant

ii. **The Ratio Test**: If $a_n > 0, \forall n \in \mathbb{N}$, we consider the quotient $\frac{a_{n+1}}{a_n}$ and we compare this quotient to 1. Then, we have:

$$\text{If } \tfrac{a_{n+1}}{a_n} > 1 \Rightarrow a_n \nearrow$$
$$\text{If } \tfrac{a_{n+1}}{a_n} < 1 \Rightarrow a_n \searrow$$
$$\text{If } \tfrac{a_{n+1}}{a_n} = 1 \Rightarrow a_n \text{ is constant}$$

iii. **The Derivative Test**: If f is a real function such that $f(n) = a_n, \forall n \in \mathbb{N}$, then we evaluate the first *derivative* of f, denoted by f', and we have that:

$$\text{If } f'(x) > 0 \Rightarrow a_n \nearrow$$
$$\text{If } f'(x) < 0 \Rightarrow a_n \searrow$$
$$\text{If } f'(x) = 0 \Rightarrow a_n \text{ is constant}$$

Example 2.5.3. Let $a_n = \frac{n+1}{n+2}$.

i. We will use the difference test to determine whether a_n is monotonic or not. We have:

$$a_{n+1} - a_n = \frac{n+2}{n+3} - \frac{n+1}{n+2} = \frac{(n+2)^2 - (n+1)\cdot(n+3)}{(n+3)(n+2)} = 1$$

Since now $a_{n+1} - a_n > 0$, we have that $a_n \nearrow$.

ii. We now use the ratio test and we have:

$$0 < \frac{a_{n+1}}{a_n} = \frac{\frac{n+2}{n+3}}{\frac{n+1}{n+2}} = \frac{(n+2)^2}{(n+1)\cdot(n+3)} < \frac{(n+1)^2}{(n+1)\cdot(n+3)} =$$
$$= \frac{n+1}{n+3} < 1 \Rightarrow a_n \nearrow$$

iii. Finally, we use the derivative test. Let $f(x) = \frac{x+1}{x+2}$. This function is a rational function and thus differentiable in its domain and in $\mathbb{R}^*_+ := (0, \infty)$ in particular. Then:

$$f'(x) = \frac{1\cdot(x+2)-(x+1)\cdot 1}{(x+2)^2} = \frac{1}{(x+2)^2} > 0 \Rightarrow$$
$$\Rightarrow f \nearrow \forall\, x \in (0, \infty) \Rightarrow a_n \nearrow$$

Exercises 2.5.4. Monotonic sequences: Using the appropriate test for monotonicity, define the sequence as strictly increasing, strictly decreasing, increasing, decreasing or constant.

1. Difference Test:

   i. $a_n = n^2 - 2n$

   ii. $a_n = n^2 - 2^n$

2. Ratio Test:

   i. $a_n = \frac{4^n}{3^n - 1}$

ii. $a_n = \frac{n!}{3^n}$

3. Derivative Test:

   i. $a_n = 1 - \frac{n}{n+1}$

   ii. $a_n = \arctan^{-1}(n)$

4. Use any test:

   i. $a_n = n^3 - 3n$

   ii. $a_n = \frac{(n+1)!}{(n-1)!}$

2.6 Limit of a sequence

We now turn our attention to the most important concept of the theory of sequences, that is, convergence or divergence. This is equivalent to studying the end behavior of a real function (the reader should be familiar with the concept of limits from courses such as "Precalculus" or "Calculus I"). In general, the notation $\lim_{n\to\infty} a_n := \lim a_n$ corresponds to evaluating the behavior of a_n as n becomes arbitrarily large.

2.6.1 Formal definition

The following (informal) definition of a limit of a sequence is very similar to the definition of the limit of a real function at infinity.

Definition 2.6.1. A sequence a_n has the limit L, denoted by $\lim a_n = L$ or $a_n \to L$ as $n \to \infty$, if we can make the terms a_n as close as we like to L by taking n *sufficiently* large. Moreover, we say that:

a_n *converges* (or is *convergent*) if $\exists \lim a_n$ and is a real number.

a_n *diverges* (or is *divergent*) if $\nexists \lim a_n$ or $\lim a_n = \infty$.

The informal definition of a sequence at infinity makes more sense if we think about it in the following way: $\lim a_n = L \in \mathbb{R}$ if as n becomes arbitrarily large, the sequence a_n *stabilizes* around L. This leads to the following *formal definition* of a limit of a sequence:

Definition 2.6.2. A sequence $(a_n)_\mathbb{N}$ *converges* to $L \in \mathbb{R}$ if-f:

$$\forall \epsilon > 0 \, \exists \, n_0 \in \mathbb{N} : |a_n - L| < \epsilon \, \forall \, n \geq n_0.$$

A sequence $(a_n)_\mathbb{N}$ *diverges* if-f:

$$\text{either} \quad \nexists \lim a_n \quad \text{or} \quad \forall \, M > 0 \, \exists \, N \in \mathbb{N} : a_n > M \, \forall \, n \geq N.$$

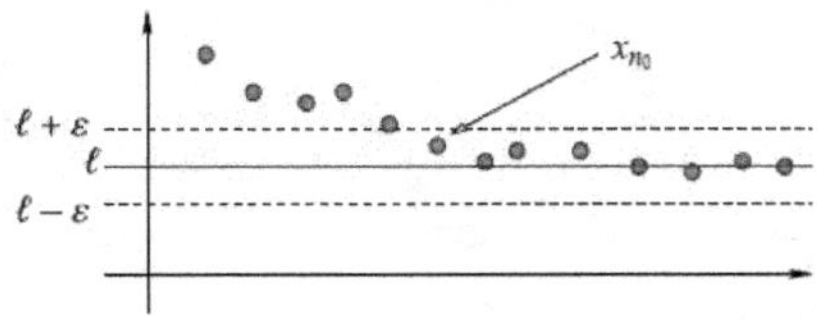

Figure 2.2: Illustration of Definition 2.6.2

Example 2.6.3. i. We prove that the sequence $a_n = \frac{3n-1}{4n+5}$ converges to $\frac{3}{4}$ using the formal definition. We have that:

$$\left|\frac{3n-1}{4n+5} - \frac{3}{4}\right| < \epsilon \Rightarrow \left|\frac{-19}{4\cdot(4n+5)}\right| < \epsilon \Rightarrow \frac{4\cdot(4n+5)}{19} > \frac{1}{\epsilon}$$

$$\Rightarrow n > \frac{1}{4}\cdot\left(\frac{19}{4\cdot\epsilon} - 5\right).$$

Choose now $n_0 = \left\lfloor \frac{1}{4}\cdot\frac{19}{4\cdot\epsilon} - 5 \right\rfloor + 1$. Then $|a_n - 3/4| < \epsilon, \forall n \geq n_0 \Rightarrow a_n \to 3/4$.

ii. We now prove that the sequence $a_n = (-1)^n$ diverges using the formal definition. Assume that $(-1)^n$ converges to a real number l. Then, for all $\epsilon > 0$, there exist $n_0 \in \mathbb{N}$ such that $|a_n - l| < \epsilon$, for $n \geq n_0$. Choose now $n_1 > n_0$, where n_1 is an even number and $n_2 > n_0$, where n_2 is an odd number. Then:

$$|(-1)^{n_1} - l| < \epsilon \Rightarrow |1 - l| < \epsilon \text{ and } |(-1)^{n_2} - l| < \epsilon \Rightarrow |-1-l| < \epsilon.$$

Since both inequalities are true for all $\epsilon > 0$, choose $\epsilon = 1/2$. Then:

$$|1-l| < 1/2 \quad \Rightarrow \quad 1/2 < l < 3/2$$

$$|-1-l| < 1/2 \quad \Rightarrow \quad -3/2 < l < -1/2$$

Obviously this is a contradiction and hence, a_n diverges.

iii. We prove that the sequence $a_n = \frac{1}{n}$ converges to 0. Indeed, let $\epsilon > 0$. Then $\frac{1}{\epsilon} > 0 \overset{Archimedean}{\underset{property}{\Rightarrow}} \exists\, n_0 \in \mathbb{N} : n_0 > \frac{1}{\epsilon} \Rightarrow n > \frac{1}{\epsilon}, \forall n \geq n_0$. We now have that:

$$n > \frac{1}{\epsilon} \Rightarrow \frac{1}{n} < \epsilon, \forall n \geq n_0 \Rightarrow$$

$$\left|\frac{1}{n}\right| < \epsilon, \forall n \geq n_0 \Rightarrow$$

$$|a_n - 0| < \epsilon, \forall n \geq n_0 \Rightarrow a_n \to 0$$

iv. The sequence $a_n = \begin{cases} n, & n \leq 5^{20} \\ \frac{1}{n}, & n > 5^{20} \end{cases}$ converges for the same reasons.

A consequence of the definition of convergence is that $a_n \to l \in \mathbb{R}$ is equivalent to $|a_n - l| \to 0$.

Out of all the sequences that diverge, the sequences that go to infinity as n becomes arbitrarily large form a specific class that is worth studying in detail.

Definition 2.6.4. We say that a sequence *diverges* to ∞ and write $a_n \to \infty$ if for any real number a, $a_n > a$ as n becomes arbitrarily large.

2.6.2 Basic Properties

Using the formal definition of a limit of a sequence we now prove some properties of the limit and we emphasize on the proofs, since the arguments used in the proofs play an important role for the solutions of the exercises that follow.

Proposition 2.6.5. *If $a_n \to l \in \mathbb{R}$, then l is unique.*

Proof. Assume that l is not unique and let $a_n \to l \neq m \leftarrow a_n$. Then:

$$a_n \to l \quad \overset{\text{Defn.}}{\Rightarrow} \quad \forall \epsilon > 0\, \exists\, n_1 \in \mathbb{N} \,:\, |a_n - l| < \epsilon,\ \forall n \geq n_1 \quad \text{and}$$

$$a_n \to m \quad \overset{\text{Defn.}}{\Rightarrow} \quad \forall \epsilon > 0\, \exists\, n_2 \in \mathbb{N} \,:\, |a_n - m| < \epsilon,\ \forall n \geq n_2.$$

We now consider an $\epsilon > 0$ and set $n_0 = \max\{n_1, n_2\}$. We have that $|a_n - l| < \epsilon$ and $|a_n - m| < \epsilon, \forall\, n \geq n_0$. Since now $l \neq m$, $|l - m| > 0$. We observe that:

$$|l - m| \;=\; |l - a_n + a_n - m| \;\leq\; |a_n - l| \;+\; |a_n - m| \;<\; 2\epsilon.$$

Since this is true for all $\epsilon > 0$, it should be true for $\epsilon = |l - m|/3 > 0$. Then, $|l - m| < \frac{2|l-m|}{3}$, that is, $1 < 2/3$ which is a contradiction. Thus, $l = m$. □

Proposition 2.6.6. *If a_n converges, then a_n is bounded.*

Proof. Let $a_n \to l \in \mathbb{R}$ and set $\epsilon = 1$. Then,

$$\exists\, n_0 \in \mathbb{N} \,:\, |a_n - l| < 1,\ \forall n \geq n_0 \Rightarrow\ l - 1 < a_n < l + 1 \quad (*)$$

Set now $M = \max\{|l - 1|, |l + 1|, a_1, a_2, \ldots, a_{n_0 - 1}\}$. Clearly:

$$|a_1| < M,\ |a_2| < M,\ \ldots,\ |a_{n_0-1}| < M \quad (**)$$

Since $|l + 1| < M \Rightarrow\ l + 1 < M$ and $|l - 1| < M \Rightarrow\ -M < l - 1$, from Eq.$(*)$ we have that

$$-M \leq a_n \leq M,\ \forall n \geq n_0 \qquad (***)$$

From Equations $(**)$ and $(***)$ we get: $|a_n| \leq M, \forall n \in \mathbb{N}$ and thus, a_n is bounded. □

Remark 2.6.7. Note that the converse of Proposition 2.6.6 is not true. For example, the sequence $(-1)^n$ is bounded $(-1 \leq (-1)^n \leq 1, \forall n \in \mathbb{N})$, but it diverges (see Example 2.6.3 ii).

We are now in position to state an important tool that we use in order to check whether a sequence a_n converges or not, by not using the formal definition of the limit. More precisely:

Theorem 2.6.8 (Test for Divergence). *If a sequence a_n is not bounded, then it diverges.*

Proof. The proof follows from Proposition 2.6.6. □

Example 2.6.9. The sequence $a_n = \frac{n^2}{3}$ is obviously not bounded and hence it diverges.

Proposition 2.6.10. *Let a_n be a convergent sequence. Then*

If $a_n \to l > 0$, then $\exists\, n_0 \in \mathbb{N} \,:\, a_n > 0,\ \forall n \geq n_0$.

If $a_n \to l < 0$, then $\exists\, n_0 \in \mathbb{N} \,:\, a_n < 0,\ \forall n \geq n_0$.

Proof. Let $a_n \to l > 0$. Set $\epsilon = l/2 > 0$. Then:

$$\exists n_0 \in \mathbb{N} \,:\, |a_n - l| < \frac{l}{2},\ \forall n \geq n_0 \Rightarrow 0 < \frac{l}{2} < a_n < \frac{3l}{2},\ \forall n \geq n_0.$$

Similarly, for $l < 0$, set $\epsilon = -l$ and get $a_n < \frac{l}{2} < 0$. □

It is worth mentioning here that the converse of Proposition 2.6.10 is not true in general. Consider for example the sequence $a_n = \frac{1}{n}$. Although $a_n > 0, \forall n \in \mathbb{N}$, $a_n \to 0$.We have the following result:

Proposition 2.6.11. *If a_n converges to l and*

$$\exists\, n_1 \in \mathbb{N} \,:\, a_n > 0, \forall n \geq n_1,$$

then $l \geq 0$.

Proof. Let $l > 0$. Then, from Proposition 2.6.10 we have that

$$\exists\, n_0 \in \mathbb{N} \,:\, a_n < 0, \forall n \geq n_0.$$

Set $n_2 = \max\{n_0, n_1\}$. Then $\forall n \geq n_0, a_n < 0$ and $a_n > 0$, which is a contradiction. Thus, $l \geq 0$. □

2.6.3 The algebra of limits

Theorem 2.6.12 (Limit Laws for Sequences). *If $a_n \to l_1 \in \mathbb{R}$, $b_n \to l_2 \in \mathbb{R}$ and $c \in \mathbb{R}$ is constant, then:*

i. $a_n + b_n \to l_1 + l_2$.

ii. $a_n - b_n \to l_1 - l_2$.

iii. $c \cdot a_n \to c \cdot l_1$, *and* $c \to c$.

iv. $a_n \cdot b_n \to l_1 \cdot l_2$.

v. $\frac{a_n}{b_n} \to \frac{l_1}{l_2}$, *if* $l_2 \neq 0$.

vi. $a_n^m \to l_1^m$, *if* $m > 0$ *and* $a_n > 0, \forall n$.

Proof. We only prove (i) and (iii). The rest follow similarly and are left as exercises.

Since $a_n \to l_1 \in \mathbb{R}$, $b_n \to l_2 \in \mathbb{R}$, we have that

$$\forall \epsilon > 0,\ \exists\, n_0 \text{ and } n_1 \in \mathbb{N} : \begin{cases} |a_n - l_1| < \epsilon/2, & \forall n \geq n_0 \\ |b_n - l_2| < \epsilon/2, & \forall n \geq n_1 \end{cases}$$

i. Set $n_2 = \max\{n_0, n_1\}$. Then

$$|a_n + b_n - l_1 - l_2| \leq |a_n - l_1| + |b_n - l_2| < 2 \cdot \frac{\epsilon}{2} = \epsilon.$$

iii. Since $a_n \to l_1 \in \mathbb{R}$, a_n is bounded, i.e. $\exists M > 0 : |a_n| \leq M$, for all $n \in \mathbb{N}$. We now have:

$$\begin{aligned} |a_n b_n - l_1 l_2| &= |a_n b_n - a_n l_2 + a_n l_2 - l_1 l_2| \\ &\leq |a_n b_n - a_n l_2| + |a_n l_2 - l_1 l_2| = \\ &= |a_n| \cdot |b_n - l_2| \ + \ |a_n - l_1| \cdot |l_2| \leq \\ &\leq M \cdot |b_n - l_2| \ + \ |a_n - l_1| \cdot |l_2| \end{aligned}$$

Since now $b_n \to l_2 \in \mathbb{R}$,

$$\exists\, n_1 \in \mathbb{N} \ : \ |b_n - l_2| < \epsilon/2M, \forall n \geq n_1$$

and similarly,

$$\exists\, n_2 \in \mathbb{N} \ : \ |a_n - l_1| < \epsilon/2|l_2|, \forall n \geq n_2.$$

Set $n_0 = \max\{n_1, n_2\}$, then for $n \geq n_0$:

$$|a_n b_n - l_1 l_2| \ \leq \ M \cdot |b_n - l_2| \ + \ |a_n - l_1| \cdot |l_2| \ < \ \epsilon \ \overset{\text{Defn.}}{\Rightarrow}$$

$$a_n \cdot b_n \to l_1 \cdot l_2.$$

□

It is worth mentioning that the converse of these limit laws are not true in general. Consider for example the sequences $(-1)^n$ and $(-1)^{n+1}$. Although both sequences diverge, their product converges since

$$(-1)^n \cdot (-1)^{n+1} = -1 \to -1.$$

Exercises 2.6.13. Convergent / Divergent Sequences

1. Prove that $\frac{1}{\sqrt{n}} \to 0$ using the definition.
2. Let $0 < a < 1$, prove that $a^n \to 0$.
3. Prove that $\frac{1}{n^2} \to 0$ using the definition.
4. Prove that the sequence $a_n = \begin{cases} n, & n \text{ even} \\ \frac{1}{n}, & n \text{ odd} \end{cases}$ diverges.

2.7 Basic tests for convergence

Our aim is to study the end behavior of a sequence, that is, we care about evaluating the limit of a sequence as n becomes arbitrarily large (∞). In this section we present some tests for convergence (or divergence) in order to avoid the use of the formal definition of the limit.

Proposition 2.7.1 (Absolute Convergence Test). *If a_n converges to l, then $|a_n|$ converges to $|l|$. The converse is not true unless $l = 0$.*

Proof. Let $\epsilon > 0$. Then, $a_n \to l \overset{\text{Defn.}}{\Rightarrow} \exists\, n_0 \in \mathbb{N} \,:\, |a_n - l| < \epsilon, \forall n \geq n_0$. We have that:

$$||a_n| - |l|| \;<\; |a_n - l| \;<\; \epsilon,\ \forall\, n \geq n_0 \;\Rightarrow\; |a_n| \to |l|.$$

In order to prove that the converse is not true in general we provide a counter-example.

Let $a_n = (-1)^n$. We have that $|a_n| = 1 \to 1$, but a_n diverges (see Example 2.6.3 ii).

If $|a_n| \to 0$, then $\forall\, \epsilon > 0\, \exists\, n_0 \in \mathbb{N} \,:\, ||a_n| - 0| < \epsilon, \forall n \geq n_0 \;\Rightarrow\; |a_n| < \epsilon, \forall n \geq n_0 \;\Rightarrow\; a_n \to 0$. □

Example 2.7.2. Let $a_n = \frac{(-1)^n}{n}$. Then, $|a_n| = \left|\frac{(-1)^n}{n}\right| = \frac{1}{n} \to 0$. Hence, $a_n \to 0$ (see Example 2.6.3 iii).

Proposition 2.7.3. *Let $a_n \to 0$ and b_n a bounded sequence. Then,*

$$a_n \cdot b_n \to 0.$$

Proof. Since b_n is a bounded sequence, $\exists\, \theta > 0 \;:\; |b_n| \leq \theta,\ \forall\, n \in \mathbb{N}$. Let now $epsilon > 0$. Then, for $\frac{\epsilon}{\theta} > 0\, \exists\, n_0 \in \mathbb{N} \,:\, |a_n - 0| < \frac{\epsilon}{\theta},\ \forall\, n \geq n_0$. Then:

$$|a_n \cdot b_n| \;=\; |a_n| \cdot |b_n| \;\leq |a_n| \cdot \theta \;<\; \frac{\epsilon}{\theta} \cdot \theta \;=\; \epsilon \Rightarrow a_n \cdot b_n \to 0.$$

□

Example 2.7.4. Let $a_n = \frac{\sin n}{n}$. Then:

$$-1 \leq \sin n \leq 1 \Rightarrow \frac{-1}{n} \leq \frac{\sin n}{n} \leq \frac{1}{n}.$$

Since now $\frac{-1}{n} \to 0 \leftarrow \frac{-1}{n}$, we have that $\frac{\sin n}{n} \to 0$.

The previous proposition can be generalized to the following (well-known from entry level calculus courses) theorem, the *Squeeze Theorem*.

Theorem 2.7.5 (Squeeze Theorem). *Let a_n, b_n, c_n be three sequences such that $b_n \leq a_n \leq c_n, \forall n \geq n_1$. If $b_n \to l \leftarrow c_n$, then $a_n \to l$.*

Proof. Let $\epsilon > 0$. Then:

$$b_n \to l \;\Rightarrow\; \exists\, n_2 \in \mathbb{N} : |b_n - l| < \epsilon,\ \forall\, n \geq n_2 \;\Rightarrow$$

$$\Rightarrow\; l - \epsilon < b_n < l + \epsilon$$

$$c_n \to l \;\Rightarrow\; \exists\, n_3 \in \mathbb{N} : |c_n - l| < \epsilon,\ \forall\, n \geq n_3 \;\Rightarrow$$

$$\Rightarrow\; l - \epsilon < c_n < l + \epsilon$$

Set $n_0 = \max\{n_1, n_2, n_3\}$. Then, $\forall n \geq n_0$ we have that:

$$l - \epsilon < b_n < a_n < c_n < l + \epsilon \Rightarrow l - \epsilon < a_n < l + \epsilon$$

$$\Rightarrow |a_n - l| < \epsilon \Rightarrow a_n \to l.$$

□

Example 2.7.6. Let $a_n = \frac{\cos n}{n^2}$. Then,

$$-1 \leq \cos n \leq 1, \forall\, n \in \mathbb{N} \Rightarrow \frac{-1}{n^2} \leq \frac{\cos n}{n^2} \leq \frac{1}{n^2},$$

and since $\frac{-1}{n^2} \to 0 \leftarrow \frac{1}{n^2}$, we conclude that $\frac{\cos n}{n^2} \to 0$.

Theorem 2.7.7. *If a_n is a bounded and monotonic sequence, then it converges. In particular:*

i. If $a_n \nearrow$ and is bounded above, then $a_n \to \sup\{a_n\}$.

ii. If $a_n \searrow$ and is bounded below, then $a_n \to \inf\{a_n\}$.

Proof. We only prove the case where a_n is bounded above and increasing. The other case follows similarly.

Let $\epsilon > 0$ and set $S = \sup\{a_n : n \in \mathbb{N}\}$. Then, from the property that characterizes the supremum we have that

$$\exists\, n_0 \in \mathbb{N} : a_{n_0} + \epsilon > S \;\Rightarrow\; a_{n_0} > S - \epsilon,$$

and since $a_n \nearrow$, we have that

$$a_n \geq a_{n_0}, \forall\, n \geq n_0.$$

Thus, $a_n > S - \epsilon, \forall\, n \geq n_0$.
Moreover, since $a_n \leq S, \forall\, n \in \mathbb{N}$, we have that

$$a_n < S + \epsilon, \forall\, n \geq n_0.$$

Thus:

$$S - \epsilon < a_n < S + \epsilon, \forall\, n \geq n_0 \;\Rightarrow |a_n - S| < \epsilon, \forall\, n \geq n_0 \;\Rightarrow\; a_n \to S.$$

□

Example 2.7.8. Let $a_n = \frac{1}{n}$. Then a_n is strictly decreasing (*why?*) and bounded below. Moreover, $\inf\{a_n\} = 0$ and hence, $\frac{1}{n} \to 0$. Note that the converse is not true. Consider for example the sequence $\frac{(-1)^n}{n}$. This sequence converges but is not monotonic.

2.7.1 del'Hospital Rules

Let now $x_0 \in \mathbb{R} \bigcup \{-\infty, +\infty\}$ and f, b be two real functions such that

$$f(n) = a_n, g(n) = b_n, \ \forall\, n \in \mathbb{N}.$$

Moreover, let f and g be differentiable functions, such that:

$$\lim_{x \to x_0} f(x) \ = \ \lim_{x \to x_0} g(x) \ = \ 0 \text{ or } \lim_{x \to x_0} f(x) \ = \ \lim_{x \to x_0} g(x) \ = \ \pm\infty$$

and $\lim\limits_{x \to x_0} \frac{f'(x)}{g'(x)}$ exists and is equal to a real number. Then:

$$\lim \frac{a_n}{b_n} \ = \ \lim \frac{f(x)}{g(x)} \ = \ \lim \frac{f'(x)}{g'(x)} \tag{2.1}$$

Equations 2.1 are known as **deL' Hospital rules**, shorthanded to DLH, (or Bernoulli rules to do justice to Bernoulli who actually came up with the rules and proved them) and they play an important role in the evaluation of limits of sequences. The proof of deL'Hospital rules is omitted.

Example 2.7.9. Let $a_n = n \cdot e^{-n}$. Then, the function $f(x) = x \cdot e^{-x}$ coincides with a_n when we restrict its domain to $\mathbb{N}$. Hence:

$$\lim a_n \ = \ \lim_{x \to \infty} \frac{x}{e^x} \ \overset{DLH}{=} \ \lim_{x \to \infty} \frac{1}{e^x} \ = \ 0.$$

Exercises 2.7.10. Test for Convergence

1. Determine whether the following sequences converge or diverge.

   i. $a_n = \frac{1}{n!}$

   ii. $a_n = \frac{(-1)^{n+1} \cdot n}{n^2+1}$

   iii. $a_n = \frac{1+\sin(2n)}{\sqrt{1+n}}$

   iv. $a_n = \frac{2n+3}{n-2}$

   v. $a_1 = 1 \ \& \ a_{n+1} = \sqrt{a_n}, \forall n \in \mathbb{N}$

   vi. $a_n = \frac{3n^2-5n}{4n^2-4n+e}$

   vii. $a_n = \sqrt{n+1} - \sqrt{n}$

2. If $b_n \to b \neq 0$, prove that $\exists\ N : |b_n| > \frac{1}{2}|b|, \forall n > N$.

3. If $a_n \to l$, prove that $\frac{a_1+a_2+\ldots a_n}{n} \to l$.

4. Prove that $\frac{a^n}{n!} \to 0, \forall a \in \mathbb{R}$.

5. Use the formal definition of the limit to prove that $2^{-1/\sqrt{n}} \to 1$.

6. Let $A \subset \mathbb{R}$ such that $A \neq \emptyset$. Then, $\exists\ a_n, b_n$ sequences such that $a_n \to \sup A$ and $b_n \to \inf A$.

7. If $a_n \geq l, \forall n \in \mathbb{N}$ and $a_n \to 0$, prove that $\sqrt{a_n} \to \sqrt{l}$.

2.7.2 Some Basic Limits

We now present some basic and important sequences and their limits.

A. $a^{1/n} \to 1$, for all $a > 0$.

We consider three cases. For $a = 1$ it is obvious that $1^{1/n} \to 1$.

Let $a > 1$. Then:

$$a^{1/n} > 1,\ \forall n \in \mathbb{N} \Rightarrow \exists\, \theta_n > 0 : a^{1/n} = 1 + \theta_n, \forall\, n \in \mathbb{N}.$$

Moreover:

$$a = (1+\theta_n)^n, \forall\, n \in \mathbb{N} \overset{\text{Bernoulli's}}{\underset{\text{inequality}}{\Rightarrow}} (1+\theta_n)^n \geq 1 + n \cdot \theta_n.$$

Now: $1 + n \cdot \theta_n > n \cdot \theta_n \Rightarrow a > n \cdot \theta_n$ and thus:

$$0 < \theta_n < a \cdot \frac{1}{n},\ \forall n \in \mathbb{N}.$$

From the Squeeze Theorem we get that $\theta_n \to 0$ and hence:

$$a^{1/n} = 1 + \theta_n \to 1.$$

Let now $0 < a < 1$. Then, $\frac{1}{a} > 1 \Rightarrow \left(\frac{1}{a}\right)^{1/n} \to 1 \Rightarrow a^{1/n} \to 1$.

B. $n^{1/n} \to 1$, for all $n \in \mathbb{N}$.

For $n > 2$ we have that: $n^{1/n} > 1 \Rightarrow a_n = n^{1/n} - 1 > 0$. We now use the *Binomial Theorem* [1] and we have that:

$$n = (1+a_n)^n = 1 + na_n + \frac{n(n-1)}{2}a_n^2 + \ldots + a_n^n \geq \frac{n(n-1)}{2} \cdot a_n^2$$

That is, $0 < a_n < \sqrt{\frac{2}{n-1}} \to 0 \overset{a_n \to 0}{\Rightarrow} (n)^{1/n} \to 1$.

C. $a_n = \left(1 + \frac{1}{n}\right)^n \to e$, the *Euler number*.

It is left as an exercise to prove that the sequence $b_n = \left(1+\frac{1}{n}\right)^{n+1}$ is a decreasing sequence. Then, through b_n we prove that a_n is increasing. More precisely:

$$a_n > b_{n-1} \cdot \left(\tfrac{n-1}{n}\right) = \left(\tfrac{n}{n-1}\right)^{n-1} = \left(1 + \tfrac{1}{n-1}\right)^{n-1} =$$

$$= a_{n-1}$$

Moreover, the sequences a_n and b_n are bounded and in particular: $2 < a_n, b_n < 3, \forall n$, and since $\frac{a_n}{b_n} = a + \frac{1}{n} \to 1$, a_n and b_n both converge to a number on $(2, 3)$. Finally one needs to show that $2 < a_n \leq e \leq b_n < 3$, and thus $a_n \to e \leftarrow b_n$.

[1] Binomial Theorem: Let x, y be two variables and let $n \in \mathbb{N}$. Then:

$$(x+y)^n = \sum_{k=0}^{n} \binom{n}{k} \cdot x^{n-k} y^k$$

D. The sequence $\{a_n\}_{n\in\mathbb{N}}$ defined recursively by

$$a_1 = 1 \text{ and } a_{n+1} = \sqrt{1+a_n}$$

converges to $\frac{1+\sqrt{5}}{2}$.

It is left as an exercise to prove that $1 \le a_n \le 2$ using induction and that a_n is increasing. Thus, a_n converges.

Let $a_n \to l \in [1,2]$. Then,

$$\lim a_{n+1} = l \Rightarrow l^2 = \lim a_{n+1}^2 = \lim(1+a_n) = 1+l$$

$$\Rightarrow l^2 - l - 1 = 0 \Rightarrow l = \tfrac{1+\sqrt{5}}{2}.$$

Exercises 2.7.11. Review Exercises on Convergence

1. If $a_n \ge 0$ and $a_n \to a > 0$, then ${a_n}^{1/n} \to 1$.
2. $a^n \to 0$, for all $a \in \mathbb{R} : |a| < 1$.
3. If $a_n > 0, \forall n$ and $\frac{1}{a_n} \to 0$, then $a_n \to +\infty$.
4. If $a_n \to \pm\infty$, then $\frac{1}{a_n} \to 0$.
5. If $a_n \to l$, then $\frac{a_1+a_2+\ldots a_n}{n} \to l$.

2.8 The Ratio & the Root Tests

In this subsection we present two very important tests for convergence, the Ratio and the nth Root test. The ratio test is due to d'Alembert and the root test is due to Cauchy. These two tests are very strong and we use them especially when a sequence contains factorials (ratio test) or powers (root test). We begin with the ratio test:

Theorem 2.8.1 (The Ratio Test; d'Alembert). *Let a_n be a sequence of real numbers such that $a_n > 0, \forall n \in \mathbb{N}$ and let $\frac{a_{n+1}}{a_n} \to \lambda$ Then:*

- *If $\lambda < 1$, then $a_n \to 0$.*
- *If $\lambda > 1$, then a_n diverges.*
- *If $\lambda = 1$, then the test is inconclusive.*

A stronger version of the ratio test that involves sequences of real numbers in general and not necessarily of positive terms is given below:

Theorem 2.8.2 (The Ratio Test; d'Alembert). *Let a_n be a sequence of real numbers and let $\left|\frac{a_{n+1}}{a_n}\right| \to \lambda$ Then:*

- *If $|\lambda| < 1$, then $a_n \to 0$.*
- *If $|\lambda| > 1$, then a_n diverges.*

- *If* $|\lambda| = 1$, *then the test is inconclusive.*

Proof. **First method:** Set $b_n = |a_n|$. We will prove first that

$$\text{if } \exists\, n_0 \,:\, b_{n+1} \le \lambda b_n, \forall\, n \ge n_0, \text{ where } 0 \le \lambda < 1, \text{ then } b_n \to 0.$$

Indeed we have:

$$0 \le b_{n+n_0} \le \lambda \cdot b_{n+n_0-1} \le \lambda^2 \cdot b_{n+n_0-2} \le \ldots \le \lambda^n \cdot b_{n_0}.$$

Since now $0 \le \lambda < 1$, we have that $\lambda^n \to 0$ and thus,

$$\lambda^n \cdot b_{n_0} \to 0 \Rightarrow b_n \to 0 \Rightarrow a_n \to 0.$$

Moreover,

$$\left|\frac{a_{n+1}}{a_n}\right| \to \lambda < 1 \Rightarrow \exists \mu \,:\, \lambda < \mu < 1$$

and

$$\exists\, k \in \mathbb{N} \,:\, \frac{b_{n+1}}{b_n} < \mu, \forall n \ge k.$$

Thus, $b_n \to 0$.

Second method: Let a_n be a sequence. Consider the sequence $b_n = |a_n|$. Then $b_n > 0, \forall n \in \mathbb{N}$, and so the ratio test (Theorem 2.8.1) may be applied. If $\frac{b_{n+1}}{b_n} = \lambda < 1 \Rightarrow b_n \to 0 \overset{b_n=|a_n|}{\Rightarrow} a_n \to 0$. Same for $\lambda = 1$ and $\lambda > 1$. □

Example 2.8.3. i. Let $a_n = \frac{a^n}{n!}$ for $a > 0$. We have that $a_n > 0$, for all $n \in \mathbb{N}$ and

$$\frac{a_{n+1}}{a_n} = \frac{\frac{a^{n+1}}{(n+1)!}}{\frac{a^n}{n!}} = \frac{a^{n+1} \cdot n!}{a^n \cdot (n+1)!} = \frac{a}{n+1} \to 0 \Rightarrow a_n \to 0.$$

ii. Let $a_n = \frac{n^n}{n!}$. We have that $a_n > 0, \forall n \in \mathbb{N}$ and

$$\frac{a_{n+1}}{a_n} = \frac{\frac{(n+1)^{n+1}}{(n+1)!}}{\frac{n^n}{n!}} = \frac{(n+1)^n \cdot (n+1) \cdot n!}{(n+1) \cdot n! \cdot n^n} =$$

$$= \frac{(n+1)^n}{n^n} = \left(1 + \frac{1}{n}\right)^n \to e > 1 \Rightarrow a_n \text{ diverges.}$$

Remark 2.8.4. In order to use the ratio test we first check whether the sequence a_n consists of positive terms only and if not, we consider the absolute value of a_n. Then, we set up the quotient $\frac{a_{n+1}}{a_n}$ (or $\left|\frac{a_{n+1}}{a_n}\right|$), we simplify it as much as we can and finally, we consider its limit as n tends to infinity. We emphasize on this because the most common mistake that students make when using the Ratio Test is that they forget to consider the limit of the quotient $\frac{a_{n+1}}{a_n}$.

As mentioned in the introduction, the Ratio Test is one very strong test for convergence of sequences and because of that, one may think that this test may replace all other tests we reviewed earlier. Unfortunately, when a sequence is defined by a "simple" formula, this test usually fails. For example, if $a_n = 1/n$, then $\frac{a_{n+1}}{a_n} = \frac{n}{n+1} \to 1$ and thus, the ratio test is inconclusive.

We now turn our focus on the nth Root Test (or just root test) by Cauchy. This test together with the ratio test are considered the most powerful tests for convergence of both sequences and series. We omit the proof of the root test in the case of sequences and we will prove it in the case of Series in the next chapter.

Theorem 2.8.5 (The Root Test; Cauchy). *Let a_n be a sequence of real numbers such that $a_n > 0, \forall n \in \mathbb{N}$ and let $a_n^{1/n} \to \lambda$ Then:*

- *If $\lambda < 1$, then $a_n \to 0$.*
- *If $\lambda > 1$, then a_n diverges.*
- *If $\lambda = 1$, then the test is inconclusive.*

As in the case of the ratio test, a stronger version of the root test that involves sequences of real numbers in general and not necessarily of positive terms is given below:

Theorem 2.8.6 (The Root Test; Cauchy). *Let a_n be a sequence of real numbers and let $\left|a_n^{1/n}\right| \to \lambda$ Then:*

- *If $|\lambda| < 1$, then $a_n \to 0$.*
- *If $|\lambda| > 1$, then a_n diverges.*
- *If $|\lambda| = 1$, then the test is inconclusive.*

The proof of this theorem is left as an exercise.

Example 2.8.7. Let $a_n = \left(\frac{n+1}{n^2}\right)^n$. Then, $a_n > 0, \forall n \in \mathbb{N}$ and

$$a_n^{1/n} = \frac{n+1}{n^2} \to 0 \Rightarrow a_n \to 0.$$

Exercises 2.8.8. Ratio and Root Tests

1. Determine whether the following sequences converge or diverge using the Ratio or the Root test:

   i. $a_n = \frac{a^n}{n!}$, for all a.

   ii. $a_n = \frac{a^n}{n^k}$, for $a > 1$ and $k \in \mathbb{N}$.

   iii. $a_n = \frac{n}{2^n}$.

   iv. $a_n = \frac{n!(n+1)!}{(2n)!}$.

   v. $a_n = \frac{(-12)^n}{n}$.

   vi. $a_n = \left(\frac{n+1}{n^2+2}\right)^n$.

   vii. $a_n = \frac{1}{e^n}$.

2.9 Sub-sequences & the Bolzano-Weierstrass Theorem

In this subsection we focus on a special operation defined on sequences and which is equivalent to the operation of composition of two real functions.

Definition 2.9.1. Let a_n, k_n be two sequences such that k_n is strictly increasing. Then we define the sequence a_{k_n} that we call a *subsequence* of a_n.

As in the case of real functions where a real function can be considered as a composition of itself with x, a sequence a_n is a subsequence of itself, by considering $n_k = k$. All other subsequences of a_n are obtained by "deleting" terms from a_n in a specific way defined by k_n. This is the reason why we call it a subsequence (and not a composite sequence). In particular, if $a_n = \{a_1, a_2, \ldots, a_n, \ldots\} = A$, then

$$a_{k_n} = \{a_{k_1}, a_{k_2}, \ldots, a_{k_n}, \ldots\} \subset A.$$

Example 2.9.2. i. Let $a_n = n + 1$ and $k_n = n^2$. Obviously k_n is strictly increasing and thus, we can form the subsequence $b_n = a_{k_n}$:

$$\begin{array}{lll} \text{For } n = 1: & k_1 = 1 & b_1 = a_1 = 2 \\ \text{For } n = 2: & k_2 = 4 & b_2 = a_4 = 5 \\ \text{For } n = 3: & k_3 = 9 & b_3 = a_9 = 10 \\ \vdots & & \\ \text{For } n: & k_n = n^2 & b_n = a_{n^2} = n^2 + 1 \end{array}$$

ii. Other examples of subsequences of a sequence a_n is the subsequence of even indices and the subsequence of odd indices

$$a_2, a_4, \ldots = a_{2n}, \qquad a_1, a_3, \ldots = a_{2n-1}$$

These subsequences are of great importance as we shall see later in this subsection.

Proposition 2.9.3. *Let $a_n \to l \in \mathbb{R}$. Then, $a_{k_n} \to l$, for all subsequences a_{k_n}.*

Proof.

$$a_n \to l \in \mathbb{R} \;\Rightarrow\; \forall\, \epsilon > 0\, \exists\, n_0 \,:\, |a_n - l| < \epsilon,\ \forall n \geq n_0 \;\Rightarrow$$

$$|a_{k_n} - l| < \epsilon,\ \forall k_n \geq k_{n_0} \geq n_0. \text{ So, } a_{k_n} \to l$$

□

Note that in the proof we used the fact that if k_n is strictly increasing and $k_n > 0,\ \forall n \in \mathbb{N}$, then $k_n \geq n,\ \forall n \in \mathbb{N}$ (the proof is based on induction and is left as an exrcise).

An immediate consequence of Proposition 2.9.3 is the following result:

Corollary 2.9.4 (Divergence Criterion). *If a_{k_n} and a_{m_n} are two subsequences of a_n such that either $a_{k_n} \to l_1 \neq l_2 \leftarrow a_{m_n}$, or one of them diverges, then a_n diverges.*

Example 2.9.5. Consider the sequence $a_n = (-1)^n$. Then,

$$a_{2n} = 1 \to 1 \neq -1 \leftarrow -1 = a_{2n-1} \text{ and thus } a_n \text{ diverges.}$$

Note that Proposition 2.9.3 can not be used in order to prove that a sequence converges. For that reason we have the following proposition which is a much stronger result.

Proposition 2.9.6. *Let a_n be a sequence. If $a_{2n} \to l \leftarrow a_{2n-1}$, then $a_n \to l$.*

Proof. Let $\epsilon > 0$. Then:

$$\begin{aligned} a_{2n} \to l \quad &\Rightarrow \exists\, n_1 \in \mathbb{N} : |a_{2n} - l| < \epsilon,\ \forall n \geq n_1 \\ a_{2n-1} \to l \quad &\Rightarrow \exists\, n_2 \in \mathbb{N} : |a_{2n-1} - l| < \epsilon,\ \forall n \geq n_2 \end{aligned}$$

Set $n_0 = \max\{n_1, n_2\}$. Then,

$$|a_{2n} - l| < \epsilon \text{ and } |a_{2n-1} - l| < \epsilon \text{ for all } n \geq n_0.$$

That is, $|a_n - l| < \epsilon$ for all $n \geq n_0 \Rightarrow a_n \to l$. □

Remark 2.9.7. Proposition 2.9.6 tells us that in order to check whether a sequence a_n converges or diverges, it suffices to consider the subsequence of even indices a_{2n}, and the subsequence of odd indices a_{2n-1}, and determine their end behavior. This behavior determines the end behavior of the sequence a_n.

Theorem 2.9.8 (Monotone Subsequence Theorem)**.** *If a_n is a sequence of real numbers, then there is a subsequence of a_n that is monotone.*

The proof is left as an exercise. The idea is that we can always "construct" a monotonic subsequence.

This leads to the following theorem, which is the main theorem of this subsection.

Theorem 2.9.9 (Bolzano-Weierstrass)**.** *If a sequence is bounded, then is has a convergent subsequence.*

Proof. Let a_n be a bounded sequence. Then, all subsequences of a_n are bounded and from Proposition 2.9.8, a_n has a monotonic subsequence. Thus, a_n contains a bounded and monotonic subsequence and hence, this subsequence converges. □

Remark 2.9.10.

i. Theorem 2.9.9 guarantees the existence of a convergent subsequence but it is sometimes difficult to find such a subsequence. For example, $a_n = \sin n$.

ii. It is also worth mentioning that if a sequence is not bounded, then we can't know whether this sequence has a convergent subsequence or not. Consider for example the sequences

$$a_n = \{0, 1, 0, 3, 0, 5, 0, \ldots\} \text{ and } b_n = \{-1, 2, -3, 4, -5, \ldots\}.$$

Obviously $a_{2n+1} = \{0, 0, \ldots\} \to 0$ and there doesn't exist a convergent subsequence of b_n.

2.10 Cauchy Sequences

Augustin-Louis Cauchy (1789-1857) was born in Paris just after the start of the French Revolution. He was very lucky to meet the mathematicians Laplace and Lagrange who were impressed by young Cauchy's mathematical talent and helped him begin his career. Cauchy made substantial contributions to many areas and published eight books and 789 papers. He was one of the most important mathematicians in the first half of the nineteenth century compared by many to the prince of mathematics, Gauss.

The definition of a convergent sequence requires us to know the limit of the sequence and that's what makes this definition weak. It is some times particularly difficult to "guess" that limit, especially when the sequence diverges. Cauchy's work towards an answer to the question of convergence is very important and useful. In particular, assume that $a_n \to l \in \mathbb{R}$. Then, for all $\epsilon > 0$ there exist an $n_0 \in \mathbb{N}$ such that $|a_n - l| < \frac{\epsilon}{2}$ for all $n \geq n_0$. Consider now two arbitrary indices n, m that are greater than n_0. Then:

$$|a_n - a_m| = |a_n - l + l - a_m| \leq |a_n - l| + |a_m - l| < \epsilon.$$

This leads to the following definition:

Definition 2.10.1. A sequence a_n is called *Cauchy*, if

$$\forall\, \epsilon > 0\, \exists\, n_0 \,:\, |a_n - a_m| < \epsilon,\ \forall\, n, m \geq n_0.$$

It is worth mentioning here that for a sequence a_n to be a Cauchy sequence, *all pairs* $\{a_n, a_m\}$ should eventually come closer and it does not suffice only $(a_{n+1} - a_n) \to 0$.

Example 2.10.2. Let $a_n = 1 + \frac{1}{2} + \ldots + \frac{1}{n}$. We have that

$$a_{n+1} - a_n \to 0, \text{ but } |a_{2n} - a_n| = \frac{1}{n+1} + \frac{1}{n+2} + \ldots + \frac{1}{2n} \geq n \cdot \frac{1}{2n} = \frac{1}{2}.$$

So, the sequence a_n is not a Cauchy sequence.

We now present Cauchy's Convergence criterion:

Theorem 2.10.3 (Cauchy's Convergence Criterion). *A sequence a_n is a Cauchy sequence if-f a_n converges.*

The proof of Theorem 2.10.3 follows from the following two Lemmas:

Lemma 2.10.4. *If a_n is a Cauchy sequence, then it is bounded.*

Proof. Let a_n be a Cauchy sequence. Then:

$$\forall\, \epsilon > 0\, \exists\, n_0 \,:\, |a_n - a_m| < \epsilon,\ \forall\, n, m \geq n_0 \Rightarrow \forall n \geq n_0 \,:\, |a_n - a_{n_0}| < \epsilon$$

Thus, $|a_n| < |a_{n_0}| + \epsilon$ and we get that the set

$$\{a_{n_0}, a_{n_0+1}, \ldots\} = \{a_n \,:\, n \geq n_0\} \text{ is bounded.}$$

Since the set $\{a_1, a_2, \ldots, a_{n_0-1}\}$ is finite, it is also bounded and thus, a_n is bounded as a union of two bounded sets. □

Lemma 2.10.5. *If a_n is a Cauchy sequence and $\exists\, a_{k_n}$ such that a_{k_n} converges to $l \in \mathbb{R}$, then $a_n \to l$.*

Proof. Let $\epsilon > 0$. Since a_n is a Cauchy sequence, we have that:

$$\exists\, n_0 \in \mathbb{N} \,:\, \forall\, n, m \geq n_0 \,:\, |a_n - a_m| < \epsilon/2.$$

Moreover, the subsequence a_{k_n} converges to l, that is,

$$\exists\, n_1 \in \mathbb{N} \,:\, \forall\, n \geq n_1 \,:\, |a_{k_n} - l| < \epsilon/2$$

and since k_n is strictly increasing,

$$\exists\, n_2 \in \mathbb{N} \,:\, k_{n_2} \geq \max\{n_0, k_{n_1}\}.$$

Consider now $n \geq n_0$. Then, $\forall n \geq n_0$ we have that

$$|a_n - l| \leq |a_n - a_{k_{n_2}}| + |a_{k_{n_2}} - l| < \epsilon \;\Rightarrow\; a_n \to l.$$

□

The importance of Theorem 2.10.3 lies in the fact that we can prove that a sequence converges without the need of evaluating its limit.

2.10.1 $\liminf$ and $\limsup$ of a sequence a_n

Definition 2.10.6. Let a_n be a bounded sequence. We define the following sequences:

$$L_n \;:=\; \inf_{k\geq n} a_k, \qquad U_n \;:=\; \sup_{k\geq n} a_k.$$

Obviously: $L_n \leq a_n \leq U_n$ and both L_n and U_n are bounded. Moreover, L_n is increasing and U_n is decreasing, and thus L_n and U_n converge. Set:

$$\liminf a_n \;:=\; \lim L_n \;:=\; \underline{\lim} a_n, \qquad \limsup a_n \;:=\; \lim U_n \;:=\; \overline{\lim} a_n$$

Lemma 2.10.7. *Let a_n be a bounded sequence and a_{k_n} a convergent subsequence. Then:*

$$\liminf a_n \;\leq \lim a_n \;\leq\; \limsup a_n.$$

Proof. Since $L_{k_n} \;\leq\; a_{k_n} \;\leq\; U_{k_n} \;\Rightarrow\; \lim L_{k_n} \;\leq\; \lim a_{k_n} \;\leq\; \lim U_{k_n}$. □

Proposition 2.10.8. *Let a_n be a bounded sequence. Then there exist subsequences a_{k_n} and a_{m_n} such that:*

$$a_{m_n} \;\to\; \liminf a_n, \text{ and } a_{k_n} \;\to\; \limsup a_n.$$

Proof. Let $u_n = \limsup a_n$. Then, since $U_n \to u$,

$$\exists\, v_1 \in \mathbb{N} \,:\, u - 1 < U_{v_1} < u + 1 \Rightarrow u - 1 \leq \sup_{k\geq v_1} < u + 1.$$

From the definition of the supremum we have that

$$\exists\, k_1 \geq v_1 \;:\; u - 1 < a_{k_1} < u + 1.$$

Similarly,

$$U_n \to u,\; \exists\, v_2 > k_1 \;:\; u - 1/2 < U_{v_2} < u + 1/2.$$

Again, $\exists\, k_2 \geq v_2 > k_1 \;:\; u - 1/2 < a_{k_2} < u + 1/2$. Continue in that way and obtain: $u - 1/n < a_{k_n} < u + 1/n \;\Rightarrow\; a_{k_n} \to u$. □

The above results lead to the following theorem:

Theorem 2.10.9. *Let a_n be a bounded sequence. Then,*

$$a_n \to l \in \mathbb{R} \Leftrightarrow \underline{\lim} a_n = \overline{\lim} a_n = l.$$

Proof. $\Rightarrow$ If $l = \pm\infty$, then it follows that $\underline{\lim} a_n = \overline{\lim} a_n = l$, from Definition 2.10.6.

If $a_n \to l \in \mathbb{R}$, then from Lemma 2.10.7, $\exists\, a_{p_n}, a_{q_n}$ such that

$$a_{p_n} \to \overline{\lim} a_n \text{ and } a_{q_n} \to \underline{\lim} a_n.$$

But since all subsequences of a_n converge to l, we get that

$$\underline{\lim} a_n = \overline{\lim} a_n = l.$$

$\Leftarrow$ If $\underline{\lim} a_n = \overline{\lim} a_n = l$, then all subsequences of a_n converge to l and thus, a_n converges to l. □

2.11 Limit points & Continuity in terms of Sequences

Definition 2.11.1. A point $l \in \mathbb{R}$ is a *limit point* of a set $A \subset \mathbb{R}$ if

$$\exists\, a_n \,:\, a_n \in A,\ \forall\, n \in \mathbb{N} \text{ such that } a_n \neq l\ \forall\, n \in \mathbb{N} \text{ and } a_n \to l.$$

The following lemma is close related to the Bolzano-Weierstrass theorem and it will be used in order to relate sequences with continuous functions. The proof is left as an exercise.

Lemma 2.11.2. *If a_n is unbounded above, then a_n has a subsequence a_{k_n} such that $a_{k_n} \to +\infty$. Similarly, if a_n is unbounded below, then $\exists\, a_{m_n} : a_{m_n} \to -\infty$.*

Before we present how sequences can be used in order to determine whether a function is continuous or not, we remind the reader the formal definition of a limit of a function and through that, the definition of *continuous* functions. Note that in introductory mathematical courses such as "Calculus I", a function is defined to be continuous on an interval I if there are no breaks in the graph of the function on I.

Definition 2.11.3. i. We say that $\lim\limits_{x \to x_0} f(x) = L$, if

$$\forall\, \epsilon > 0\, \exists\, \delta_{\epsilon,x} > 0 \,:\, |f(x) - L| < \epsilon \text{ when } |x - x_0| < \delta.$$

ii. A function f is *continuous* at a point $x_0 \in dom(f)$ if

$$\lim_{x \to x_0} f(x) = f(x_0).$$

iii. Combining (i) and (ii), we say that a function f is continuous at a point $x_0 \in dom(f)$ if

$$\forall\, \epsilon > 0\, \exists\, \delta_{\epsilon,x} > 0 \,:\, |f(x) - f(x_0)| < \epsilon \text{ when } |x - x_0| < \delta.$$

Using now Lemma 2.11.2 and Definition 2.11.3, we prove the following theorem.

Theorem 2.11.4. *Let f be a function defined on a closed interval $[a, b]$ and let $l \in [a, b]$. Then, f is continuous at l if and only $\lim\limits_{n \to +\infty} f(a_n) = f(l)$, for all sequences a_n such that $a_n \in [a, b]$ and $a_n \to l$. Moreover, if $l = a$ or $l = b$, then we consider the one-sided limits.*

Proof. We sketch the proof and consider the case where $l \in (a, b)$. The cases $l = a$ and $l = b$ follow similarly.

Let $l \in (a, b)$ and assume that f is continuous at l and a_n is a sequence with terms in $[a, b]$ and $a_n \to l$. Then,

$$\text{for } \epsilon > 0 \; \exists \, \delta > 0 \; : \; |f(x) - f(l)| < \epsilon, \text{ when } |x - l| < \delta.$$

Since now $a_n \to l$, we have that

$$\exists \, n_0 \in \mathbb{N} \; : \; |a_n - l| < \delta, \; \forall \, n \geq n_0.$$

Thus, the conditions stated above are necessary. Sufficiency of the conditions is left as an exercise. □

Review Test on Sequences

1. State whether the following statements are true or false. If false provide a counterexample.

   i. Every divergent sequence is unbounded.
   ii. Every subsequence of a divergent sequence diverges.
   iii. Every strictly increasing sequence is bounded below.
   iv. Every bounded monotone sequence is convergent.
   v. If $|a_n| \to |a|$, then $a_n \to a$.
   vi. If a_n diverges and $c \in \mathbb{R} \backslash \{0\}$, then the sequence $c \cdot a_n$ diverges.
   vii. If $a_n \to a \in \mathbb{R}$ and $b_n \to +\infty$, then $a_n \cdot b_n \to +\infty$.

2. State the definition of a convergent sequence a_n.

3. Prove that if a sequence a_n converges to $l \in \mathbb{R}$, then l is unique.

4. Prove that the sequence $a_n = (-1)^n$ diverges using the definition.

5. Prove that the sequence $a_n = \frac{2n}{1+n}$ is monotonic. Is it bounded?

6. Prove that the sequence $a_n = \frac{2^n}{2n+5}$ diverges.

7. Prove that if a sequence is strictly increasing, then it is bounded below.

8. Give an example of a bounded but divergent sequence. Explain.

9. Prove that if $a_n \to l$, then $a_{k_n} \to l$.

10. Prove that $\frac{a^n}{n!} \to 0, \forall a > 0$.

11. Prove that the sequence $a_n = \frac{3n-1}{4n+5}$ converges using the definition.

12. Prove that the sequence $a_n = \frac{1+sin(2n)}{\sqrt{n+1}}$ converges.

13. Let $0 \leq a \leq b$. Prove that $(a^n + b^n)^{1/n} \to b$.

14. Prove that a Cauchy sequence is bounded.

15. Decide whether the following sequences converge or diverge. If they converge, evaluate the limit.

   i. $a_n = n - \sqrt{n}$
   ii. $a_n = \frac{(-1)^{n-1}n}{n^2+1}$
   iii. $a_n = \frac{n+4}{n^2-1}$
   iv. $a_n = (1 + \frac{1}{2n})^n$
   v. $a_n = ln(5n^2 + 1) - ln(3n^2 + 2)$
   vi. $a_n = n^k \cdot a^n$, where $a \in (0, 1)$ and $k \in \mathbb{Z}$.

vii. $a_n = \frac{n^2-2018\cdot(n-1)}{e^{n^2+3n}}$

viii. $a_n = \frac{2n^3+n-5}{n^2-4}$

ix. $a_n = \frac{e^{n+1}+2^n}{e^n+2^{n+1}}$

x. $a_n = n \cdot \sin\left(\frac{1}{n}\right)$

xi. $a_n = \sqrt{9n^2+27n} + an$, for $a \in \mathbb{R}$

xii. $a_n = \frac{(k^2+k-2)\cdot n^2+(k-1)\cdot n+2}{(k+5)\cdot n+7}$, for $k \in \mathbb{R}$

16. Is the sequence $a_n = 1 + \frac{1}{2} + \frac{1}{3} + \ldots + \frac{1}{n}$ a Cauchy sequence?

Chapter 3

Infinite Series

In this chapter we study another type of sequences called *series*, a fundamental object in mathematics. In fact, understanding series was the motivation behind the development of the foundation of analysis. We briefly introduced series in §2.2 and §2.3, where we studied arithmetic and geometric series respectively. The reader is advised to review these sections at this time, since we will not repeat the definitions and results presented there.

3.1 Series and convergence

Roughly speaking, a series is an expression of the form:

$$\sum_{n=1}^{\infty} a_n := a_1 + a_2 + \ldots$$

Definition 3.1.1. Let a_n be a sequence. We define a new sequence

$$s_n := a_1 + a_2 + \ldots + a_n, n \in \mathbb{N},$$

the sequence of *partial sums*. Then:

$$\lim s_n := \lim \sum_{i=1}^{n} a_i := \sum_{n=1}^{\infty} a_n.$$

Definition 3.1.2. We say that the series $\sum_{n=1}^{\infty} a_n$ *converges* to $s \in \mathbb{R}$, denoted as

$$\sum_{n=1}^{\infty} a_n = s$$

if the sequence of partial sums of a_n converges to s. Moreover, the series diverges if s_n diverges.

It is worth reminding the reader here the case of **geometric series** with common ratio $r < 1$. Consider for example $a_n = \frac{1}{2^n}$. Then:

$$s_n = \frac{1}{2} \cdot \frac{1 - \frac{1}{2^n}}{1 - \frac{1}{2}} \Rightarrow \lim s_n = \lim \left(1 - \frac{1}{2^n}\right) = 1 \Rightarrow \sum_{n=1}^{\infty} \frac{1}{2^n} = 1.$$

Thus, the series $\sum_{n=1}^{\infty} \frac{1}{2^n}$ converges to 1. On the other hand, consider the sequence $a_n = (-1)^n$. Then, $\sum_{n=1}^{\infty} (-1)^n = -1 + 1 - 1 + 1 - \ldots$ and the series has no sum since s_n diverges.

Another important type of series is the **telescoping series**, that is, *series for which all terms of its partial sum are canceled except for the first and last terms* as shown in the following example:

$$\sum_{n=1}^{\infty} \frac{1}{n(n+1)} = \sum_{n=1}^{\infty} \left(\frac{1}{n} - \frac{1}{n-1} \right) = \ldots = 1 - \frac{1}{n+1} \to 1 \in \mathbb{R}.$$

In other words, $\sum_{n=1}^{\infty} a_n = s$ means that by adding sufficiently many terms of the series we can get close to the number s. But as noted above, not all series converge. In this chapter we will set up the necessary tools in order to be able to determine whether a series converges or diverges. Note also that from the properties of limits, we have that the sum of a convergent series is **unique**.

Before we proceed, we present the following result which is the analogue of the "tail" of a sequence.

Proposition 3.1.3. *Let* $\sum a_n$ *be a series. Let* $n_0 \in \mathbb{N}$. *Then*

$$\sum_{n=1}^{\infty} a_n \quad \text{converges if and only if} \quad \sum_{n=n_0}^{\infty} a_n \quad \text{converges.}$$

Proof. Consider the partial sums of the two series:

$$\sum_{n=1}^{k} a_n = \left(\sum_{n=1}^{n_0 - 1} a_n \right) + \sum_{n=n_0}^{k} a_n.$$

and note that $\sum_{n=1}^{n_0-1} a_n$ is a fixed number. The result follows. □

Exercises 3.1.4. Prove that the *harmonic series* $\sum_{n=1}^{\infty} \frac{1}{n}$ diverges.

3.1.1 Caychy Series

We begin this subsection with the Cauchy's Convergence criterion for series, since the reader should be familiar with it by now. As in the case of sequences, Cauchy series play an important role in the theory of series in general. More precisely:

Definition 3.1.5. A series $\sum a_n$ is said to be *Cauchy* or a *Cauchy series*, if the sequence of partial sums $\{s_n\}$ is a Cauchy sequence.

Recall that a sequence of real numbers converges if and only if it is Cauchy. Therefore, *a series is convergent if and only if it is Cauchy*. More precisely:

Theorem 3.1.6 (Cauchy's Convergence Criterion for Series). *A series $\sum a_n$ converges if and only*

$$\forall\, \epsilon > 0\, \exists\, n_0 \in \mathbb{N}\, :\, |a_n + a_{n+1} + \ldots + a_m| < \epsilon\, \forall m \geq n \geq n_0.$$

Proof. In terms of the partial sums s_n of $\sum a_n$,

$$s_n + a_{n+1} + \ldots + a_m = s_m - s_{n-1}.$$

Therefore,

$$|s_m - s_{n-1}| < \epsilon\, \forall\, m \geq n \geq n_0.$$

The proof is concluded since $\sum a_n$ converges if and only if s_n converges. □

An easy result of the definition of Cauchy series is the following proposition, which characterizes the Cauchy series:

Proposition 3.1.7. *The series $\sum a_n$ is Cauchy if:*

$$\forall\, \epsilon > 0,\, \exists\, n_0 \in \mathbb{N}\, :\, \forall\, k > n \geq n_0,\, \Rightarrow \left|\sum_{i=n+1}^{k} a_i\right| < \epsilon.$$

The proof of Proposition 3.1.7 is left as an exercise.

3.2 Basic properties

Proposition 3.2.1. *Let $\sum\limits_{n=1}^{\infty} a_n$ and $\sum\limits_{n=1}^{\infty} b_n$ be two convergent series and let $\lambda \in \mathbb{R}$. Then the series $\sum\limits_{n=1}^{\infty}(a_n + b_n)$ and $\sum\limits_{n=1}^{\infty}(\lambda \cdot a_n)$ converge and in particular:*

$$\sum_{n=1}^{\infty}(a_n + b_n) = \sum_{n=1}^{\infty} a_n + \sum_{n=1}^{\infty} b_n, \quad \text{and} \quad \sum_{n=1}^{\infty}(\lambda \cdot a_n) = \lambda \cdot \sum_{n=1}^{\infty} a_n.$$

Proof. Let $\sum\limits_{n=1}^{\infty} a_n = \lim s_n = a \in \mathbb{R}$, $\sum\limits_{n=1}^{\infty} b_n = \lim \sigma_n = b \in \mathbb{R}$ and $\sum\limits_{n=1}^{\infty}(a_n + b_n) = \tau_n$, $\sum\limits_{n=1}^{\infty}(\lambda \cdot a_n) = t_n$. Then:

$$\begin{aligned} \sum_{n=1}^{\infty}(a_n + b_n) &= \lim(s_n + \sigma_n) \overset{\text{seq.}}{=} \lim s_n + \lim \sigma_n = a + b = \\ &= \sum_{n=1}^{\infty} a_n + \sum_{n=1}^{\infty} b_n. \end{aligned}$$

Moreover:

$$\sum_{n=1}^{\infty}(\lambda \cdot a_n) = \lim(\lambda \cdot s_n) = \lambda \cdot \lim s_n = \lambda \cdot a = \lambda \cdot \sum_{n=1}^{\infty} a_n.$$

□

Remark 3.2.2. It is worth mentioning that if $\sum_{n=1}^{\infty} a_n$ converges and $\sum_{n=1}^{\infty} b_n$ diverges, then $\sum_{n=1}^{\infty}(a_n + b_n)$ diverges. To see that, assume that $\sum_{n=1}^{\infty}(a_n + b_n)$ converges. Then, from Proposition 3.2.1 and for $\lambda = -1$ we have that $\sum_{n=1}^{\infty}(-a_n)$ converges. Then, the series:

$$\sum_{n=1}^{\infty}[(a_n + b_n) + (-a_n)]$$

converges. Since the series $\sum_{n=1}^{\infty} b_n$ diverges, we reach at a contradiction.

Remark 3.2.3. If $\sum_{n=1}^{\infty} a_n$ and $\sum_{n=1}^{\infty} b_n$ diverge, then $\sum_{n=1}^{\infty}(a_n+b_n)$ either converge or diverge. Consider for example the sequences $a_n = (-1)^n$ and $b_n = (-1)^{n+1}$. Then, $\sum_{n=1}^{\infty} a_n$ and $\sum_{n=1}^{\infty} b_n$ diverge, but $\sum_{n=1}^{\infty}(a_n + b_n)$ converges (*why?*).

Remark 3.2.4. Multiplying series is not as simple as adding. We cannot multiply term by term, since this is not even true for finite sums:

$$(a + b)(c + d) \neq ac + bd.$$

We will deal with product of series at the end of this chapter where we will deal with the *Cauchy product* of series.

Exercise 3.2.5. Let $\sum a_n, \sum b_n$ and $\sum c_n$ be three series such that $c_n = a_n + b_n, \forall n \in \mathbb{N}$. Prove the **Three Series Principle**, that is, if any two of these three series converge, then so does the third.

We are now ready to state the first test for *divergence*.

Theorem 3.2.6 (Divergence Test). *If $\sum_{n=1}^{\infty} a_n$ converges, then $a_n \to 0$. Equivalently, if a_n does not converge to zero, then $\sum_{n=1}^{\infty} a_n$ diverges.*

Proof. Since $\sum_{n=1}^{\infty} a_n$ converges, s_n converges. Let $s_n \to s \in \mathbb{R}$. Then, $\lim s_{n-1} = s$ and thus:

$$\lim(s_n - s_{n-1}) = \lim s_n - \lim s_{n-1} = s - s = 0.$$

But since $a_n = s_n - s_{n-1}$, we have that $a_n \to 0$. □

Example 3.2.7. The series $\sum_{n=1}^{\infty} \frac{n-1}{n+2}$ diverges, since $\frac{n-1}{n+2} \to 1 \neq 0$.

Remark 3.2.8. The converse of the Divergence test is not true in general. Consider for example the sequence $a_n = \frac{1}{n} \to 0$. We observe that:

$$\begin{aligned} s_{2n} - s_n &= (a_1 + a_2 + \ldots + a_{2n}) - (a_1 + a_2 + \ldots + a_n) = \\ &= a_{n+1} + a_{n+2} + \ldots + a_{2n} = \\ &= \tfrac{1}{n+1} + \tfrac{1}{2n} \geq \tfrac{1}{2n} + \ldots + \tfrac{1}{2n} = n \cdot \tfrac{1}{2n} = \tfrac{1}{2} \end{aligned}$$

Thus $s_{2n} - s_n \geq \frac{1}{2}, \forall n \in \mathbb{N}$. Assume now that $\sum\limits_{n=1}^{\infty} \frac{1}{n}$ converges. Then s_n converges. Let $s_n \to l \in \mathbb{R}$ and thus, $s_{2n} \to l$ (subsequence). Then:

$$\begin{aligned} s_{2n} - s_n &\geq \tfrac{1}{2} \Rightarrow \lim(s_{2n} - s_n) \geq \lim \tfrac{1}{2} = \tfrac{1}{2} \Rightarrow \\ &\Rightarrow \lim s_{2n} - \lim s_n \geq \tfrac{1}{2} \Rightarrow l - l \geq \tfrac{1}{2} \Rightarrow 0 \geq 1/2, \end{aligned}$$

which is a contradiction.

The series $\sum\limits_{n=1}^{\infty} \frac{1}{n}$ is called the **harmonic series** and it will play an important role in the theory of convergent and divergent series.

Exercises 3.2.9. Divergence test for series
Determine whether the following series converge or diverge:

1. $\sum\limits_{n=1}^{\infty} (-1)^n$.

2. $\sum\limits_{n=1}^{\infty} \frac{e^n}{\ln n}$.

3. $\sum\limits_{n=1}^{\infty} n$.

3.3 The Integral Test

From now on we consider sequences that converge to zero. The test for divergence fail in these cases since $a_n \to 0$ doesn't imply that s_n converges too. Our intention is to present all the tools needed in order to be able to determine whether such series converge or not. Towards that end we have the following straightforward result:

Proposition 3.3.1. *If $\sum\limits_{n=1}^{\infty} a_n$ converges, then $s_n := a_1 + a_2 + \ldots + a_n$ is bounded.*

Proof. Since $\sum\limits_{n=1}^{\infty} a_n$ converges, we have that $\sum\limits_{n=1}^{\infty} a_n = \lim s_n = a \in \mathbb{R}$. So s_n convegres and thus, s_n is bounded. □

Remark 3.3.2. The converse of Proposition 3.3.1 is not true in general. For example, if $a_n = (-1)^n$, then s_n is bounded (*why?*), but $\sum\limits_{n=1}^{\infty} (-1)^n$ diverges.

The question that arises now is the following:

Under what restrictions can we have the converse of Proposition 3.3.1?

The answer to that question comes from the theory of sequences: *if s_n is also increasing then it converges*, and in particular we have that

$$s_n \to \sup\{s_n\} := \sum a_n.$$

It is easy to see that if we focus on sequences that consist of positive terms *only*, then s_n is increasing by definition. Thus, we have the following powerful result:

Proposition 3.3.3 (The Boundedness Criterion). *Let $a_n \geq 0, \forall n$. Then, $\sum\limits_{n=1}^{\infty} a_n$ converges if-f s_n is bounded above.*

Proof. $\Rightarrow$ If $\sum\limits_{n=1}^{\infty} a_n$ converges, then from Proposition 3.3.1 we have that s_n is bounded and thus, bounded above.

$\Leftarrow$ Let s_n be bounded above. We observe that

$$s_{n+1} - s_n = a_{n+1} \geq 0 \;\Rightarrow\; s_{n+1} \geq s_n \;\Rightarrow\; s_n \text{ is strictly increasing.}$$

Thus, s_n is strictly increasing and bounded above, hence, s_n converges.

□

Using the Boundedness criterion we can determine whether a series converges or diverges by only checking whether s_n is bounded above or not. Note that even if we can prove that s_n is bounded above, most likely we won't be able to determine its limit, i.e. $\sup\{s_n\}$. We will deal with that later.

Motivation

i. Consider the series $\sum\limits_{n=1}^{\infty} \frac{1}{n^2}$. We observe that the test for divergence fails here, since $\frac{1}{n^2} \to 0$. Moreover, $a_n \geq 0, \forall n \in \mathbb{N}$ and thus, s_n is an increasing sequence. To check whether s_n is bounded above (and thus, convergent), we use a geometric argument based on the *Riemann definition of integrals*. More precisely, let $f(x) = \frac{1}{x^2}$. Consider now *rectangles* that lie below the graph of f and whose base is equal to one (see Figure 3.3 for Riemann sums in general). Then, the areas of the rectangles are $\frac{1}{1^2}, \frac{1}{2^2}, \frac{1}{3^2}, \ldots$, exactly the terms of the series. We now have that:

$$s_n < \text{ area of the region bounded by the graph of } f \text{ and the } x - \text{axis } + 1.$$

In other words, $s_n < 1 + \int\limits_1^{\infty} \frac{1}{x^2}\,dx \,:\, \lim\limits_{t\to\infty} \int\limits_1^{t} \frac{1}{x^2}\,dx \,+1 \,=\, 2$. That means that s_n is bounded above and since it is increasing, it converges to a number $l < 2$.

Remark 3.3.4. We emphasize on the fact that $s_n \to l < 2$ and most likely $2 \neq \sup\{s_n\}$. That is, although we know that s_n converges, it is very difficult to know it's exact value. *Euler* proved that $s_n \to \frac{\pi^2}{6}$, but the proof is beyond the scope of these notes.

ii. Consider now the harmonic series $\sum\limits_{n=1}^{\infty} \frac{1}{n}$. We observe that the test for divergence fails here again, since $\frac{1}{n} \to 0$. Moreover, we have that $a_n \geq 0, \forall n \in \mathbb{N}$ and thus, s_n is an increasing sequence. Using the same geometric argument as in (i) we

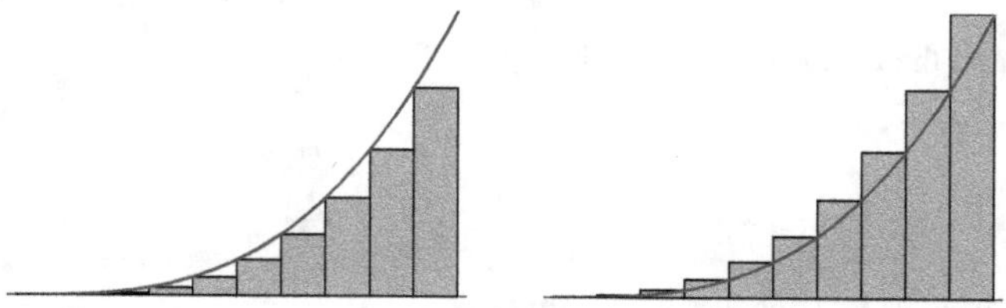

Figure 3.1: Left and Right Riemann sums

have that the areas of the rectangles that lie below the graph of f and whose base is equal to one (see Figure 2) are $\frac{1}{1}, \frac{1}{2}, \frac{1}{3}, \ldots$, exactly the terms of the series. So:
$s_n <$ area of the region bounded by the graph of f and the $x-$axis $+\ 1$.
Since the improper integral diverges and $\lim s_n < \infty$, we can't conclude. We need to slightly alter the technique presented above as follows:
Instead of considering rectangles that lie below that curve of f, C_f, consider rectangles that lie above C_f. Then:

$$s_n > \frac{1}{1} + \frac{1}{2} + \ldots + \frac{1}{n} > \int_1^{n+1} \frac{1}{x}\, dx = \ln(n+1) \to \infty.$$

That is, $s_n > 1 + \int_1^{\infty} \frac{1}{x}\, dx = +\infty$. Hence, $s_n \to \infty$, i.e. $\sum a_n$ diverges. It can be proved that $\sum_{n=1}^{\infty} \frac{1}{n} = \infty$.

The above examples lead to the following beautiful theorem due to Maclaurin (and later from Cauchy) that relates infinite series to improper integrals and it is known as the *Integral Test*. The proof of this theorem is left as an exercise and is based on the geometric arguments presented above.

Theorem 3.3.5 (The Integral Test). *Assume that f is a continuous function, positive and decreasing for $x \geq k \in \mathbb{R}$ and $f(n) = a_n, \forall n \in \mathbb{N}$. Then:*

$$\sum_{n=1}^{\infty} a_n \quad \text{and} \quad \int_1^{\infty} f(x)\, dx$$

either both converge or both diverge.

Proof. Consider the interval $I = [1, \infty)$ and a *partition* of I,

$$\bigcup_{n=1}^{\infty} [n, n+1], n \in \mathbb{N}.$$

Fix $n_0 \in \mathbb{N}$ and compare the integral $\int_1^{n_0} f(x)\, dx$ with the Right and Left Riemann sums associated to the partition $\{1, 2, \ldots, n_0\}$. Since f is decreasing, the Right Riemann sum is $\sum\limits_{n=2}^{n_0+1} a_n$ and the Left Riemann sum is $\sum\limits_{n=1}^{n_0} a_n$. That is:

$$\sum_{n=2}^{n_0+1} a_n \leq \int_{n=1}^{n_0} f(x)\, dx \leq \sum_{n=1}^{n_0} a_n.$$

The result follows by considering the limits as $n_0 \to \infty$. □

Remark 3.3.6. If $\int\limits_1^\infty f(x)\, dx$ converges, then $\sum\limits_{n=1}^{\infty} a_n$ converges but the value of the improper integral is *not* equal to the value of the series.

With the use of the Integral Test we get the following result, which involves a family of series that is very important in the theory of series in general, the **p-series**.

Proposition 3.3.7 (The p-Series Test)**.** *The series*

$$\sum \frac{1}{n^p} \begin{cases} \text{converges,} & \text{for } p > 1 \\ \text{diverges,} & \text{for } p \leq 1 \end{cases}$$

The proof is an immediate result of the Integral Test and is left as an exercise for the reader.

Remark 3.3.8. Note that if the series $\sum \frac{1}{n^p}$ is treated as a function of p, one obtains the *Riemann ζ function*, the behavior of which remains a mystery and is one of the most famous unsolved problems in mathematics today.

Example 3.3.9. Let $a_n = \frac{n}{e^n}, n \in \mathbb{N}$ and $f(x) = \frac{x}{e^x}, x \in \mathbb{R}$. The test for divergence fails since using DLH we get $a_n \to 0$. We now check whether the conditions of the Integral Test are satisfied:

i. f is continuous as a combination of continuous functions.

ii. f is positive.

iii. f is decreasing for $x \geq 1$. To see this we need to evaluate the first derivative of f. Again, f is differentiable as a combination of differentiable functions and $f'(x) = \frac{1-x}{e^x}$. Thus,

$$f'(x) < 0 \Leftrightarrow x > 1.$$

All conditions of the Integral test are satisfied and hence we know that $\sum\limits_{n=1}^{\infty} \frac{n}{e^n}$ and $\int\limits_1^\infty \frac{x}{e^x}\, dx$ behave in the exact same way. We evaluate the improper integral and we have:

$$\int_1^\infty \frac{x}{e^x}\, dx := \lim_{t\to\infty} \int_1^t \frac{x}{e^x}\, dx \overset{IbP}{=} \lim_{t\to\infty} \left(-\frac{t+1}{e^t} + \frac{2}{e^2}\right) \overset{DLH}{=} \frac{2}{e^2} \in \mathbb{R}$$

We conclude that the series $\sum\limits_{n=1}^{\infty} \frac{n}{e^n}$ converges.

Note that in Example 3.3.9 *IbP* stands for "*Integration by Parts*" and "*DLH*" stands for *de L'Hospital rules*. The reader is advised to review some basic integration techniques at this point and especially *Integration by Parts & u-Substitution* and *improper integrals*.

3.3.1 Estimating the sum of a series

Assume that using the Integral Test we are able to show that a series $\sum a_n$ is convergent and we now want to find an approximation to the sum s of the series. Any partial sum s_n is an approximation and obviously, the approximation becomes better as n becomes larger. To get a feel of how good an approximation of s is using s_n for a fixed $n \in \mathbb{N}$, we set:

$$R_n = s - s_n$$

which is called the *remainder* or the *error* in the approximation. It is obvious from the proof of Theorem 3.3.5 that:

$$\int_{n+1}^{\infty} f(x)\,dx \leq R_n \leq \int_{n}^{\infty} f(x)\,dx.$$

This is called the **The Remainder Estimate of the Integral Test** and it is used to estimate the error in an approximation. In particular, we have the following theorem:

Theorem 3.3.10. *Assume that f is a continuous function, positive and decreasing for $x \geq k \in \mathbb{R}$ and $f(n) = a_n, \forall n \in \mathbb{N}$. If $R_n = s - s_n$, then:*

$$s_n + \int_{n+1}^{\infty} f(x)\,dx \leq s \leq s_n + \int_{n}^{\infty} f(x)\,dx. \tag{3.1}$$

Equation 3.1 provides and upper and a lower bound for s, the sum of the series $\sum a_n$.

3.4 Comparison Tests

Consider now the series $\sum\limits_{n=1}^{\infty} \frac{3^n}{4^n+a}$, where $a \in \mathbb{R}$. We observe that when $a = 0$ this is a geometric series and thus, it converges to $\frac{3}{4} \cdot \frac{1}{1-3/4} = 3$. But when $a \neq 0$, this series resembles a geometric series and we suspect that it should converge too. When $a > 0$ we have that $\frac{3^n}{4^n+a} < \frac{3^n}{4^n}$, and hence, the series $\sum\limits_{n=1}^{\infty} \frac{3^n}{4^n+a}$ must converge (*why?*). On the other hand, if $a < 0$, then $\frac{3^n}{4^n+a} > \frac{3^n}{4^n}$ and we can't conclude. In this paragraph we present two *comparison tests* that are very useful in the theory of series. The readers who are familiar with the comparison tests for improper integrals should consider these tests as their analogue. It is also worth mentioning that these tests are due to a famous and important mathematician, Bernoulli.

3.4.1 The Comparison Test

If we know about convergence of a certain series we can take advantage of the following Comparison Test to determine whether a another series converges or not. In particular:

Theorem 3.4.1 (The Comparison Test). *Let $\sum\limits_{n=1}^{\infty} a_n$ and $\sum\limits_{n=1}^{\infty} b_n$ be series with positive terms. Then:*

i. If $0 < a_n \leq b_n$ *and* $\sum_{n=1}^{\infty} b_n$ *converges, then* $\sum_{n=1}^{\infty} a_n$ *converges.*

ii. If $0 < a_n \leq b_n$ *and* $\sum_{n=1}^{\infty} a_n$ *diverges, then* $\sum_{n=1}^{\infty} b_n$ *diverges.*

Proof. Let $\sum_{n=1}^{\infty} a_n = \lim s_n$ and $\sum_{n=1}^{\infty} b_n = \lim \sigma_n$. Since $a_n \leq b_n$ we have that $s_n \leq \sigma_n$.

i. If $\sum_{n=1}^{\infty} b_n$ converges, then σ_n is bounded and since $s_n \leq \sigma_n$, s_n is bounded. Thus, from Proposition 3.3.3, the Boundedness criterion, s_n converges.

ii. Let now $\sum_{n=1}^{\infty} a_n$ be divergent and assume that $\sum_{n=1}^{\infty} b_n$ converges too. Then, from (i) $\sum_{n=1}^{\infty} a_n$ should converge. We reached at a contradiction and thus, $\sum_{n=1}^{\infty} b_n$ diverges.

□

Remark 3.4.2. In using the Comparison Test we must have some known series $\sum_{n=1}^{\infty} b_n$ for the purpose of comparison. The most commonly used series for this purpose are the **p-series** and the **geometric series**. It is worth mentioning that the Comparison Test is very useful in the cases of series of the form

$$\sum_{n=1}^{\infty} \frac{k_m n^m + k_{m-1} n^{m-1} + \ldots + k_1 n + k_0}{d_v n^v + d_{v-1} n^{v-1} + \ldots + d_1 v + d_0},$$

where $k_i, d_i \in \mathbb{R}$, $\forall\, i$ and $m, v \in \mathbb{N}$. We consider the series $\sum_{n=1}^{\infty} p \cdot \frac{1}{n^{m-v}}$ and compare $\frac{1}{n^{m-v}}$ to $\frac{k_m n^m + k_{m-1} n^{m-1} + \ldots + k_1 n + k_0}{d_v n^v + d_{v-1} n^{v-1} + \ldots + d_1 v + d_0}$ when possible.

Example 3.4.3. Consider the series $\sum_{n=1}^{\infty} \frac{n-2}{n^2+5}$. We observe that

$$\frac{n-2}{n^2+5} < \frac{n}{n^2} = \frac{1}{n},$$

and since $\sum_{n=1}^{\infty} \frac{1}{n}$ diverges (harmonic series), the series $\sum_{n=1}^{\infty} \frac{n-2}{n^2+5}$ diverges.

It is worth mentioning that the comparison test does not tell us where the series converges to. It only tell us that a limit of the partial sums exists.

Example 3.4.4. Consider the series $\sum_{n=1}^{\infty} \frac{3^{n+1}}{2n(5^n)}$. We observe that

$$\frac{3^{n+1}}{2n(5^n)} < \left(\frac{3}{5}\right)^n,$$

and since $\sum_{n=1}^{\infty} \left(\frac{3}{5}\right)^n$ converges (geometric series), the series $\sum_{n=1}^{\infty} \frac{3^{n+1}}{2n(5^n)}$ converges too.

It is worth mentioning here that although the condition $0 < a_n \le b_n$ in the Comparison Test is given for all n, we only need to verify that there exists an $n_0 \in \mathbb{N}$ such that $0 < a_n \le b_n$ for $n \ge n_0$. This is because $\sum_{n=1}^{\infty} a_n = \sum_{n=1}^{n_0-1} a_n + \sum_{n=n_0}^{\infty} a_n$, i.e. the convergence of a series is not affected by a finite number of terms (remember here the "tails" of sequences).

Remark 3.4.5. Assume that $\sum_{n=1}^{\infty} b_n$ converges and $0 < a_n \le b_n$. Using the Comparison Test we have that $\sum_{n=1}^{\infty} a_n$ converges. Let also

$$R_n := s - s_n, \quad \text{where } s_n \to s = \sum_{n=1}^{\infty} a_n$$

$$Q_n := t - t_n, \quad \text{where } t_n \to t = \sum_{n=1}^{\infty} b_n$$

If we used the Integral Test to determine that $\sum_{n=1}^{\infty} b_n$ converges, then we have an upper and a lower bound for the remainder and the error in the approximation. If b_n is a geometric sequence, then we can have the exact value of $\sum_{n=1}^{\infty} b_n$. For all cases $R_n \le Q_n$, since $a_n \le b_n$.

Example 3.4.6. Consider now the series $\sum \frac{1}{n^2+1}$. We will show that it converges. Indeed, observe that $\frac{1}{n^2+1} < \frac{1}{n^2}, \forall n \in \mathbb{N}$. Then, the series $\sum \frac{1}{n^2}$ converges (p-series, where $p = 2 > 1$) and thus, by the Comparison Test, $\sum \frac{1}{n^2+1}$ converges.

Notice now that in order to be able to apply the Comparison Test, we need to be able to prove that $a_n \le b_n,\ \forall\, n \in \mathbb{N}$. This is sometimes very challenging (or even impossible). The following test is a *stronger version* of the Comparison Test since it *weakens* this condition a lot. More precisely:

Theorem 3.4.7 (The Delayed Comparison Test). *Let $\sum a_n, \sum b_n$ be two series such that $a_n, b_n \ge 0, \forall n \in \mathbb{N}$ and $\exists\, n_0 \ :\ a_n \le b_n,\ \forall\, n \ge n_0$, then:*

$$\sum_{n=1}^{\infty} a_n \le \sum_{n=1}^{n_0} (a_n - b_n) + \sum_{n=1}^{\infty} b_n$$

Theorem 3.4.7 is called **The Delayed Comparison Test** because it only requires $a_n \le b_n$ for all *sufficiently large* n's. The proof of the Delayed Comparison Theorem is left as an exercise.

Exercises 3.4.8. The Comparison Test

1. Use the Comparison Test to determine whether the following series converge or diverge:

   i. $\sum_{n=1}^{\infty} \frac{2n}{n^3+n^2+1}$

ii. $\sum_{n=1}^{\infty} \frac{1}{n^2-2}$

iii. $\sum_{n=1}^{\infty} \frac{\sqrt{n+1}}{n^2+1}$

3.4.2 The Limit Comparison Test

We now turn our focus on the example in the introduction of this paragraph, $\sum_{n=1}^{\infty} \frac{3^n}{4^n+a}$, and the case where $a < 0$. Note that the Comparison Test fails in this case since $\frac{3^n}{4^n+a} > \frac{3^n}{4^n}$ and $\sum_{n=1}^{\infty} \frac{3^n}{4^n}$ converges. Nonetheless our intuition is that the series should converge since they look very much alike. In those cases where the Comparison Test fails, we use the Limit Comparison Test:

Theorem 3.4.9 (The Limit Comparison Test). *Let $\sum_{n=1}^{\infty} a_n$ and $\sum_{n=1}^{\infty} b_n$ be series with positive terms and let $\lim \frac{a_n}{b_n} = L$. Then:*

i. If $0 < L < +\infty$, then $\sum_{n=1}^{\infty} a_n$ and $\sum_{n=1}^{\infty} b_n$ either both converge or both diverge.

ii. If $L = 0$ and $\sum_{n=1}^{\infty} b_n$ converges, then $\sum_{n=1}^{\infty} a_n$ converges.

iii. If $L = +\infty$ and $\sum_{n=1}^{\infty} b_n$ diverges, then $\sum_{n=1}^{\infty} a_n$ diverges.

The proof of Theorem 3.4.9 is left as an exercise.

Example 3.4.10. Consider the series $\sum_{n=1}^{\infty} \frac{2n^2+2n}{\sqrt{2+n^5}}$. We observe that the series $\sum_{n=1}^{\infty} \frac{2}{n^{1/2}}$ that diverges, since it's a p-series where $p = 1/2 < 1$. Moreover, $\frac{a_n}{b_n} \to 1 > 0 \Rightarrow$ the series $\sum_{n=1}^{\infty} \frac{2n^2+2n}{\sqrt{2+n^5}}$ diverges.

Exercises 3.4.11. The Limit Comparison Test

1. Use the Limit Comparison Test to determine whether the following series converge or diverge:

   1. $\sum_{n=1}^{\infty} \frac{2^n}{3^n+1}$

   2. $\sum_{n=1}^{\infty} \frac{2018n^2-2019n-2020}{n^3+n^2+n+1}$

   3. $\sum_{n=1}^{\infty} \frac{\sqrt{n+1}}{n^2+1}$

3.5 Alternating Series

Before turning our attention to series with arbitrary terms, i.e. not necessarily positive, we first study a very important family of series, that of *alternating series*, whose terms alternate in sign.

Definition 3.5.1. An *alternating* series is a series whose terms are alternately positive and negative.

Remark 3.5.2. Alternating series are of the form $\sum(-1)^n \cdot a_n$, where $a_n > 0, \forall n \in \mathbb{N}$.

The following test for convergence of alternating series is due to *Leibniz* and we will refer to it as the *Leibniz Test* or *Alternating Series Test* and it is very powerful:

Proposition 3.5.3 (Alternating Series Test). *Let $\{a_n\}$ be a monotone decreasing sequence of positive real numbers such that $a_n \to 0$. Then*

$$\sum_{n=1}^{\infty}(-1)^n a_n$$

converges.

Proof. Let $s_m := \sum_{n=1}^{m}(-1)^n a_n$ be the mth partial sum. Then:

$$\begin{aligned} s_{2n} &= \sum_{n=1}^{2n}(-1)^n a_n = (-a_1 + a_2) + \cdots + (-a_{2n-1} + a_{2n}) = \\ &= \sum_{n=1}^{n}(-a_{2n-1} + a_{2n}). \end{aligned}$$

The sequence $\{a_n\}$ is decreasing and thus $(-a_{2n-1} + a_{2n}) \leq 0$ for all n. Therefore, the subsequence $\{s_{2n}\}$ of partial sums is a decreasing sequence.
Similarly, $(a_{2n} - a_{2n+1}) \geq 0$, and so

$$s_{2n} = -a_1 + (a_2 - a_3) + \cdots + (a_{2n-2} - a_{2n-1}) + a_{2n} \geq -a_1.$$

The sequence $\{s_{2n}\}$ is decreasing and bounded below, so it converges. Let $s_{2n} \to s$. We will show that $s_n \to s$. Indeed:

$$s_{2n+1} = s_{2n} + a_{2n+1}.$$

Given $\epsilon > 0$, pick $n_0 \in \mathbb{N}$ such that $|s_{2n} - s| < \frac{\epsilon}{2}$ whenever $2n \geq n_0$. Since $\lim a_n = 0$, we can consider n_0 such that $a_{2n+1} < \frac{\epsilon}{2}$ whenever $2n \geq n_0$. If $2n \geq n_0$, we have $|s_{2n} - a| < \frac{\epsilon}{2} < \epsilon$, so we just need to check the situation for s_{2n+1}:

$$|s_{2n+1} - s| = |s_{2n} - s + a_{2n+1}| \leq |s_{2n} - s + a_{2n+1}| < \frac{\epsilon}{2} + \frac{\epsilon}{2} = \epsilon.$$

Thus, $s_n \to s$. □

Example 3.5.4. Consider the alternating series $\sum_{n=1}^{+\infty}(-1)^n\frac{1}{n}$. We have that $a_n = \frac{1}{n}$ is positive for all n, a_n is strictly decreasing and $a_n \to 0$. By Leibniz's test, $\sum_{n=1}^{+\infty}(-1)^n\frac{1}{n}$ converges. This series is called **alternating harmonic series** and will play an important role to the sections that follow.

3.5.1 Estimating Sums

As mentioned earlier in the case of the Integral Test, a partial sum s_n of any convergent series can be used as an approximation to the total sum, say s. The remainder of the approximation, $R_n = s - s_n$, can be used to estimate the accuracy of the approximation. The next theorem provides an upper bound for the error involved when used for series that satisfy the conditions of the Alternating Series Test. More precisely:

Theorem 3.5.5 (Alternating Series Estimation Theorem). *Let $\{a_n\}$ be a monotone decreasing sequence of positive real numbers such that $a_n \to 0$ and let*

$$\sum_{n=1}^{\infty}(-1)^n a_n \;=\; s \in \mathbb{R}.$$

Then:

$$R_n \;:=\; |s - s_n| \;\leq\; a_{n+1}$$

Proof. We have that

$$R_n \;:=\; |s - s_n| \;=\; \left|\sum_{n=1}^{\infty}(-1)^n a_n \;-\; \sum_{n=1}^{n}(-1)^n a_n\right| \;\leq\; a_{n+1}.$$

□

Remark 3.5.6. We need to emphasize on the fact that the result of Theorem 3.5.5 does not apply on other types of series than alternating series that satisfy the conditions of Leibniz's test.

Example 3.5.7. Consider the series $\sum_{n=1}^{\infty}(-1)^n \cdot \frac{1}{n^2+1}$. Then, by the Alternating Series Test, this series converges to a real number, say s. We now use the 10th-partial sum of s_n,

$$s_{10} := a_1 + \ldots + a_{10} = -\frac{1}{2} + \frac{1}{5} - \ldots + \frac{1}{101} \;\simeq\; s.$$

The error involved in this approximation is less or equal the first neglected term in the sum, that is: $|R_n| \leq a_{11} \;=\; \frac{1}{122}$.

Exercises 3.5.8. Alternating Series

1. Determine whether the following series converge or diverge and in the case of convergent series, determine the error involved when using s_5 for approximating s:

    1. $\sum_{n=1}^{+\infty}\frac{(-1)^n \cdot 7n}{1+9n^2}$
    2. $\sum_{n=1}^{+\infty}\frac{(-1)^n \cdot 7n}{1+9n}$
    3. $\sum_{n=1}^{+\infty}\frac{(-1)^n \cdot n}{n+1}$
    4. $\sum_{n=1}^{+\infty}\frac{(-1)^n \cdot \sin n}{n}$

3.6 Absolute Convergence

So far we have studied two different families of series: series whose terms are **positive** and the **alternating series**. In the first case, we were able to force s_n to be increasing and thus, we only needed to check whether s_n was bounded above or not. We will now try to generalize the results presented in the previous sections for the case where a_n is not necessarily positive for all n, i.e. we will try to gain information for $\sum a_n$ from $\sum |a_n|$. We have the following definition first.

Definition 3.6.1. A series $\sum a_n$ *converges absolutely* if the series $\sum |a_n|$ converges.

Consider again the alternating harmonic series $\sum (-1)^n \frac{1}{n}$. As mentioned before, this series converges from the Leibniz Test. Note though that the series

$$\sum \left|(-1)^n \frac{1}{n}\right| = \sum \frac{1}{n}$$

diverges (harmonic series). On the other hand, if we consider the series $\sum (-1)^n$ that diverges (Test for Divergence), we have that

$$\sum |(-1)^n| = \sum 1 = 1 + 1 + \dots$$

diverges too. Our intention is to find a relation between absolute convergence and convergence of series, just like we did in the case of sequences. Although the examples noted above do not lead to a straight-forward conclusion, we have the following strong result:

Proposition 3.6.2. *If the series $\sum a_n$ converges absolutely, then it converges.*

Proof. Note that $|a_n| = a_n$, if $a_n > 0$ and $|a_n| = -a_n$ otherwise. That is:

$$0 \leq a_n + |a_n| \leq 2|a_n|.$$

Since now $\sum |a_n|$ converges, $\sum 2|a_n|$ converges too and from the Comparison Test we have that $\sum (a_n + |a_n|)$ converges. Now:

$$\sum a_n = \sum (a_n + |a_n| - |a_n|) = \sum (a_n + |a_n|) - \sum |a_n|$$

which converges as a difference of two convergent series. □

A different proof of Proposition 3.6.2 is now presented, using the notion of Cauchy series:

Proof. As mentioned before, a series is convergent if and only if it is Cauchy. Hence, assume that $\sum |a_n|$ is Cauchy, i.e.

$$\forall \epsilon > 0,\ \exists n_0 \in \mathbb{N} : \sum_{i=k+1}^{n} |a_i| = \left|\sum_{i=k+1}^{n} a_i\right| < \epsilon,\ \forall n > k \geq n_0.$$

We apply the *Cauchy-Schwartz* inequality now, i.e. the *triangle inequality*, for a finite sum to obtain:

$$\left|\sum_{i=k+1}^{n} a_i\right| \leq \sum_{i=k+1}^{n} |a_i| < \epsilon.$$

Hence, $\sum a_n$ is Cauchy and therefore it converges. □

Remark 3.6.3. The converse of Proposition 3.6.2 is not true.The alternating harmonic series $\sum_{n=1}^{\infty}(-1)^n \cdot \frac{1}{n}$ that we mentioned in the introduction of this paragraph comprise a counter example.

This remark leads to the following definition:

Definition 3.6.4. If a series converges, but does not converge absolutely, we say it is *conditionally convergent.*

The following example indicates the importance of absolute convergent series.

Example 3.6.5. The series $\sum_{n=1}^{\infty} \frac{\sin n}{n^3}$ consists of positive and negative terms that not necessarily alternate in sign. We observe that $\left|\frac{\sin n}{n^3}\right| \leq \frac{1}{n^3}$ and since $\sum_{n=1}^{\infty} \frac{1}{n^3}$ converges (p-series where $p = 3 > 1$), by the Comparison Test, $\sum_{n=1}^{\infty} \left|\frac{\sin n}{n^3}\right|$ converges, and thus, $\sum_{n=1}^{\infty} \frac{\sin n}{n^3}$ converges.

Exercise 3.6.6. Prove that if $\sum a_n$ and $\sum b_n$ converge absolutely, then $\sum a_n \cdot b_n$ converges absolutely.

3.7 The Ratio Test; d'Alembert

In this section we generalize the d'Alembert's Ratio Test for series.

Theorem 3.7.1 (Ratio test). *Let $\sum a_n$ be a series, $a_n \neq 0$ for all n, and* $\lim \frac{|a_{n+1}|}{|a_n|} = L \in \mathbb{R}$. *Then:*

If $L < 1$, then $\sum a_n$ converges absolutely.

If $L > 1$, then $\sum a_n$ diverges.

If $L = 1$, the test is inconclusive.

Proof. If $L > 1$, then a_n diverges (Ratio Test for sequences). Since it is a necessary condition for the convergence of series that $a_n \to 0$, we have that $\sum a_n$ diverges.

Assume now that $L < 1$. Obviously, $L \geq 0$. Choose r such that $L < r < 1$. Since $r - L > 0, \exists\, n_0 \in \mathbb{N} : \forall\, n \geq n_0 : \frac{|a_{n+1}|}{|a_n|} - L < r - L$. Therefore: $\frac{|a_{n+1}|}{|a_n|} < r$. Then, for $n > n_0$ we have that:

$$\begin{aligned} |a_n| &= |a_{n_0}| \cdot \frac{|a_{n_0+1}|}{|a_{n_0}|} \frac{|a_{n_0+2}|}{|a_{n_0+1}|} \cdots \frac{|a_n|}{|a_{n-1}|} < \\ &< |a_{n_0}| \cdot r \ldots r = |a_{n_0}| r^{n-n_0} = (|a_{n_0}| r^{-n_0}) r^n. \end{aligned}$$

For $k > n_0$ we have:

$$\begin{aligned}\sum_{n=1}^{k} |a_n| &= \left(\sum_{n=1}^{n_0} |a_n|\right) + \left(\sum_{n=n_0+1}^{k} |a_n|\right) \\ &< \left(\sum_{n=1}^{n_0} |a_n|\right) + \left(\sum_{n=n_0+1}^{k} (|a_{n_0}|r^{-n_0})r^n\right) \\ &= \left(\sum_{n=1}^{n_0} |a_n|\right) + (|a_{n_0}|r^{-n_0})\left(\sum_{n=n_0+1}^{k} r^n\right).\end{aligned}$$

The series $\sum_{n=0}^{\infty} r^n$ converges (geometric series with common ration less than one) and thus, $\sum_{n=n_0+1}^{\infty} r^n$ converges. We now have:

$$\begin{aligned}\sum_{n=1}^{k} |a_n| &< \left(\sum_{n=1}^{n_0} |a_n|\right) + (|a_{n_0}|r^{-n_0})\left(\sum_{n=n_0+1}^{k} r^n\right) \\ &\le \left(\sum_{n=1}^{n_0} |a_n|\right) + (|a_{n_0}|r^{-n_0})\left(\sum_{n=n_0+1}^{\infty} r^n\right).\end{aligned}$$

We conclude that the sequence of partial sums of $\sum |a_n|$ is bounded and that $\sum |a_n|$ is convergent. Thus, $\sum a_n$ is absolutely convergent. □

Example 3.7.2. Consider the series $\sum_{n=1}^{\infty} \frac{3^n}{n!}$. Then,

$$\lim \frac{\frac{3^{(n+1)}}{(n+1)!}}{\frac{3^n}{n!}} = \lim \frac{3^n \cdot 3 \cdot n!}{3^n \cdot (n+1) \cdot n!} = \lim \frac{3}{n+1} = 0.$$

By the Ratio Test the series converges absolutely.

Remark 3.7.3. As already mentioned in the previous paragraph, if a series converges absolutely, then it converges. Thus, using the Ratio Test for $\sum |a_n|$, if $L < 1$ then $\sum a_n$ converges. What makes the Ratio Test very powerful, is the fact that the rule is valid for $L > 1$ too. In general, if $\sum |a_n|$ diverges, then we can not determine whether $\sum a_n$ converges or diverges (remember the definition of a conditional convergent series). If though we prove that $\sum |a_n|$ diverges using the Ratio Test, then the series $\sum a_n$ also diverges. This result is extremely useful, since it comprises the only tool we have to attack analogous problems. The same fact holds for the *Root Test* that we present in the following paragraph.

Exercises 3.7.4. The Ratio Test

1. Determine whether the following series converge or diverge.

   i. $\sum_{n=1}^{\infty} \frac{1}{2n+1}$

ii. $\sum_{n=1}^{\infty}(-1)^n \cdot \frac{1}{2^n}$

iii. $\sum_{n=1}^{\infty}(-1)^n \cdot \frac{1}{n^3}$

2. Prove that if $\sum_{n=1}^{\infty} a_n$ converges, then $\sum_{n=1}^{\infty}(a_{2n} + a_{2n+1})$ also converges.

3.8 The Root Test; Cauchy

Theorem 3.8.1 (Root test; Cauchy). *Let $\sum a_n$ be a series and let*

$$\lim |a_n|^{1/n} = L.$$

Then:

If $L < 1$ then $\sum a_n$ converges absolutely.

If $L > 1$ then $\sum a_n$ diverges.

If $L = 1$, the test is inconclusive.

Remark 3.8.2. The root test is very helpful when a series includes powers of n and the Ratio Test is useful when the series includes factorials. The Root Test can also be applied in the presence of factorials, but it is not immediate and it is difficult to evaluate $\lim_{n\to\infty} (n!)^{1/n}$.

It is worth mentioning that the Root Test and the Ratio Test are extremely powerful tools for determining whether a series converges or diverges. In general, when the Ratio Test succeeds in determining whether a series converges or diverges, the Root Test succeeds too. What is very interesting is that in theory, the Root Test is more powerful than the Ratio Test, but the Ratio Test is easier to apply. In particular we have the following:

Theorem 3.8.3. *Let $\{a_n\}$ be a sequence of non-negative terms which is decreasing and $a_n \to 0$. Then:*

$$\text{if} \quad \lim \frac{a_{n+1}}{a_n} = L \in [0, \infty], \quad \text{then} \quad \lim (a_n)^{1/n} = L.$$

We omit the proof that is a bit technical, since it is beyond the scope of these lecture notes.

3.9 Rearrangements

As mentioned earlier, infinite sums do not behave in general as finite sums. In particular, if we *rearrange* the order of the terms in an infinite sum that converges to a real number, s, then the value of the "*new*" sum does not necessarily converge to s. In this paragraph we present how infinite sums differ from finite sums and determine in which cases these sums behave in the same way.

Definition 3.9.1. By a *rearrangement* of an infinite series $\sum a_n$ we mean a series obtained by changing the order of the terms of $\sum a_n$. More precisely, if $f : \mathbb{N} \to \mathbb{N}$ is a *bijection*, then the corresponding rearrangement is the series $\sum a_{f(n)}$.

We now have the following theorem that states that absolute convergent series behave as finite sums up to any rearrangement, due to *Weierstrass*.

Theorem 3.9.2 (Weierstrass). *Let $\sum a_n$ be an absolutely convergent series such that:*

$$\sum a_n = s \in \mathbb{R} \quad \text{and let} \quad f : \mathbb{N} \to \mathbb{N} \quad \text{be a bijection.}$$

Then $\sum a_{f(n)}$ is absolutely convergent and $\sum a_{f(n)} = s$.

We omit the proof of Theorem 3.9.2 and proceed with an example to demonstrate how powerful and useful it is.

Example 3.9.3. Consider the alternating harmonic series $\sum \frac{(-1)^n}{n}$, which does not converge absolutely. It can be shown that $\sum \frac{(-1)^n}{n} = \ln 2$. Multiply now both sides by $\frac{1}{2}$ and obtain $\sum \frac{(-1)^n}{2n} = \frac{\ln 2}{2}$. Consider the series:

$$-\frac{1}{2} + \frac{1}{4} - \frac{1}{8} + \ldots = \frac{\ln 2}{2}$$

$$0 - \frac{1}{2} + 0 + \frac{1}{4} + 0 - \frac{1}{8} + \ldots = \frac{\ln 2}{2}$$

$$\overset{(+)}{\Rightarrow} 1 + \frac{1}{3} - \frac{1}{2} + \frac{1}{5} + \ldots = \frac{3 \ln 2}{2}$$

Note that the last sum is nothing but a rearrangement of the first sum, that is:

$$\ln 2 = \sum \frac{(-1)^n}{n} = \sum \frac{(-1)^n}{2n} = \frac{3 \ln 2}{2}$$

This example demonstrates how different infinite sums behave differently than finite sums. One can say that the terms of an absolute convergent series "commute", just like real numbers under addition do. To generalize and conclude the theory of rearrangements of infinite sums, we present a result by *Riemann*:

Theorem 3.9.4 (Riemann Rearrangement Theorem). *If $\sum a_n$ is a conditionally convergent series and $a \in [-\infty, +\infty]$, then there is a bijection f such that $\sum a_{f(n)} = a$.*

From Theorem 3.9.4 we have that given **any** real number (including infinities), *there exist* a rearrangement such that the series converge to that number. Using now the result of Riemann, one may obtain a beautiful theory of *conditionally & unconditionally convergent* series, where an unconditionally convergent series is a series that converges and *any* rearrangement converges to the same sum. Then, as in the case of absolute convergence of series, one may prove the following result:

A covergent series is unconditionally convergent if-f it is absolutely convergent.

The proof of this result is an immediate result of the definitions and the Riemann and Weierstrass Theorems and the curious reader should be able to fill in the gaps (exercise).

We conclude this paragraph by stating two important results:

Theorem 3.9.5. *Let $\sum a_n$ be an absolutely convergent series. If the series of the positive terms of a_n converge to l_1 and the series of the negative terms of a_n converges to l_2, then $\sum a_n \to l_1 - l_2$.*

Exercise 3.9.6. Apply Theorem 3.9.5 for $\sum \frac{(-1)^{n-1}}{2^n}$.

Theorem 3.9.7. *If the series $\sum a_n$ is conditionally convergent, then the series of the positive terms of a_n and the series of the negative terms of a_n diverge.*

3.10 Various Tests

In this section we present some important tests for convergence of series which cannot be easily found in a textbook all together. We omit the proofs of these Tests and only present some examples to show their importance. The proofs of some tests presented here are left as exercises for the interested readers.

Theorem 3.10.1 (Cauchy Condensation Test). *Let a_n be a non-negative and decreasing sequence. Then, the series $\sum a_n$ converges if and only if $\sum 2^n a_{2^n}$ converges.*

Proof. We have that

$$\begin{aligned}\sum_{n=1}^{\infty} a_n &= a_1 + a_2 + a_3 + \ldots \leq a_1 + 2a_2 + 4a_4 + \ldots = \\ &= \sum_{n=0}^{\infty} 2^n a_{2^n} \leq 2 \cdot \sum_{n=1}^{\infty} a_n.\end{aligned}$$

The result follows by the Comparison Test. □

Example 3.10.2. Consider the series $\sum \frac{1}{n \ln n}$. Then,

$$\sum 2^n \frac{1}{2^n \ln 2^n} = \sum \frac{1}{n \ln 2} = \frac{1}{\ln 2} \sum \frac{1}{n},$$

which diverges (harmonic series). Thus, the series $\sum \frac{1}{n \ln n}$ diverges.

Exercise 3.10.3. Prove that the series $\sum_{n=2}^{\infty} \frac{1}{e^{\ln^2 n}}$, $n \geq 2$ converges.

Remark 3.10.4. The series

$$\sum \frac{1}{n \cdot \ln^p n}, \sum \frac{1}{n \cdot \ln n \cdot \ln^p(\ln n)}$$

are called *Bertrand series* of order 2 and 3 and they converge for $p > 1$. For $p \leq 1$ the Bertrand series diverge.

Theorem 3.10.5 (Catalan's logarithm Test). *Let a_n be a positive sequence and*

$$\lim \frac{\ln \frac{1}{a_n}}{\ln n} = l \in \mathbb{R}.$$

Then the series $\sum a_n$ converges if $l > 1$, it diverges when $0 \leq l < 1$ and for $l = 1$ the test is inconclusive.

Example 3.10.6. Consider the series $\sum n^{\ln a}$, for $a > 0$. Then, using Catalan's test we have that

$$\lim \frac{\ln\left(\frac{1}{n^{\ln a}}\right)}{\ln n} = -\ln a,$$

and thus the series converges for $a < 1/e$, diverges for $1/e < a \leq 1$ and for $a = \frac{1}{e}$ we obtain $\sum a_n = \sum \frac{1}{n}$, which diverges.

Remember now that the ratio test fails when the ratio $\frac{|a_{n+1}|}{|a_n|}$ tends to 1. *Kummer's Tests* provide a collection of ratio tests and is a generalization of d'Alemberts test. More precisely:

Theorem 3.10.7 (Kummer's Test). *The series $\sum a_n$ can be tested by the following criteria: Let $\{D_n\}$ denote any sequence of positive numbers and compute*

$$L = \lim_{n\to\infty} \inf \left[D_n \cdot \frac{a_n}{a_{n+1}} - D_{n+1}\right].$$

If $L > 0$, the series $\sum a_n$ converges.

If $D_n \cdot \frac{a_n}{a_{n+1}} - D_{n+1} \leq 0$ for sufficiently large values of n, then if the series $\sum \frac{1}{D_n}$ diverges, then the series $\sum a_n$ diverges.

An immediate result of this theorem is that Kummer's test coincides with d'Alembert's Ratio Test when $D_n = 1$ for all n.

Another variant on the ratio test is known as *Raabe's test*:

Theorem 3.10.8 (Raabe's Test). *The series $\sum a_n$ can be tested by the following criterion:*

$$L = \lim_{n\to\infty} n \cdot \left[\frac{a_n}{a_{n+1}} - 1\right].$$

If $L < 1$, the series $\sum a_n$ converges.

If $L > 1$, the series $\sum a_n$ diverges.

If $L = 1$, the test is inconclusive.

Exercise 3.10.9. Use Raabe's Test to prove that the series

$$\sum \frac{1 \cdot 3 \cdot 5 \cdot \ldots \cdot (2n-1)}{4 \cdot 6 \cdot 8 \cdot \ldots \cdot (2n+2)}$$

converges.

The following test generalizes Raabe's test:

Theorem 3.10.10 (Generalization of Raabe's Test). *Let $\sum a_n$ be a series with positive terms. Let also:*

$$L = \lim_{n\to\infty} n \cdot \left[n \cdot \left(\frac{a_n}{a_{n+1}} - 1\right) - 1\right] \cdot \ln n$$

If $L > 1$, the series $\sum a_n$ converges.

If $L < 1$, the series $\sum a_n$ diverges.

If $L = 1$, the test is inconclusive.

Exercise 3.10.11. Prove that the series $\sum\limits_{n=1}^{\infty} \frac{\ln n}{n}$ diverges.

A test that generalizes, in a sense, Leibniz's Test is the following test due to *Abel*:

Theorem 3.10.12 (Abel's Test)**.** *Assume that $\sum a_n$ is a series whose partial sums form a bounded sequence and let $\{b_n\}$ be a sequence such that $b_n \to 0$, and $\sum |b_{n+1} - b_n|$ converges. Then $\sum b_n \cdot a_n$ converges.*

Exercise 3.10.13. Prove that if the series $\sum\limits_{n=1}^{\infty} a_n$ converges, then so does $\sum\limits_{n=1}^{\infty} a_n \cdot \frac{1}{n}$.

Theorem 3.10.14 (Zame's Test)**.** *The series $\sum a_n$, where $a_n > 0\, \forall n \in \mathbb{N}^*$ converges if $\exists\, k > 1 \,:\, (1 - a^{1/n}) \cdot \frac{n}{\ln n} \geq k,\ \forall n \geq n_0$ and it diverges when $(1 - a^{1/n}) \cdot \frac{n}{\ln n} \leq 1$.*

Exercise 3.10.15. Prove that the series $\sum\limits_{n=1}^{\infty} \left(1 - 3 \cdot \frac{\ln n}{n}\right)^n$ converges.

A test that can be used for alternating series is the following test which is due to *Dirichlet*:

Theorem 3.10.16 (Dirichlet's Test)**.** *Assume that $\sum a_n$ and $\sum b_n$ are two series such that:*

The partial sums of b_n are bounded for all n.

The sequence a_n is decreasing and tends to zero.

Then, the series $\sum (a_n \cdot b_n)$ converges.

An interested reader should now combine Dirichlet's Test with Leibniz's Test and obtain what is known as the **Almost Alternating Series Test** (exercise).

Exercise 3.10.17. Prove that the series $\sum\limits_{n=1}^{\infty} (-1)^{n-1} \frac{1}{n^2}$ converges.

We conclude this subsection with *Gauss's Ratio Test*:

Theorem 3.10.18 (Gauss's Ratio Test)**.** *The series $\sum a_n$ can be tested by the following criterion: Assume that:*

$$\frac{a_n}{a_{n+1}} = 1 + \frac{L}{n} + \frac{\phi(n)}{n^2}$$

where ϕ forms a bounded sequence. Then:

If $L > 1$, the series $\sum a_n$ converges.

If $L \leq 1$, the series $\sum a_n$ diverges.

Exercises 3.10.19. Various Tests

1. Determine whether the following series converge or diverge:

   i. $\sum \frac{\ln n}{n^2}$

   ii. $\sum \frac{1}{n(\ln n)^2}$

   iii. $\sum \frac{1}{n(\ln n)}$

   iv. Use the Cauchy Condensation Test to prove that the p-series $\sum_{n=1}^{\infty} \frac{1}{n^p}$ converges for $p > 1$.

3.11 Strategy for Testing Series

Given an infinite series $\sum_{n=1}^{\infty} a_n = a_1 + a_2 + \ldots + a_n + \ldots$, here is an outline procedure to test its convergence:

i. Check whether the series is a p-series, a geometric series or a telescoping series. If not, check whether some algebraic manipulation may bring the series into one of these forms.

ii. If $\lim a_n \neq 0$, then the series diverges. If $\lim a_n = 0$, then:

iii. If the form of the series is similar to a p-series or a geometric series, use the Comparison Test. Note that you may need to check for absolute convergence.

iv. Use the Limit Comparison Test if the Comparison Test fails. If all the above fail then:

v. Use the Integral Test if its hypothesis are satisfied.

vi. If $\lim \frac{|a_{n+1}|}{|a_n|} = L$, then the series is convergent if $L < 1$, it is divergent if $L > 1$ and if $L = 1$ the ratio test fails and the result is indeterminate.

vii. If $\lim |a_n|^{1/n} = L$, then the series is convergent if $L < 1$, it is divergent if $L > 1$ and if $L = 1$ the root test fails and the result is indeterminate.

viii. Use the alternating series test if possible.

ix. Use other tests, such as the Cauchy test, Raabes test, Gauss test etc if all the above fail.

Exercises 3.11.1. Strategy & Review

1. Compute the sum of the following series.

   i. $\sum_{n=1}^{\infty} \frac{2^n+(-1)^n}{\pi^{2n}}$

   ii. $\sum_{n=1}^{\infty} \frac{2n+1}{(n^2+1)(n^2+2n+2)}$ (telescoping sum)

2. Test the following series for convergence. Briefly justify your answers.

i. $\sum_{n=1}^{\infty} \frac{1}{e^{1/n}}$

ii. $\sum_{n=0}^{\infty} \frac{\pi^n}{n!}$

iii. $\sum_{n=1}^{\infty} \frac{\sqrt{n+2}}{n^2}$

iv. $\sum_{n=1}^{\infty} \frac{n}{(2n-3)(2n-4)}$

v. $\sum_{n=2}^{\infty}(-1)^n \frac{\ln(n)}{n}$

vi. $\sum_{n=1}^{\infty}(-1)^n \left(1+\frac{1}{n}\right)^{1/n}$

3.12 The Cauchy product of Series

As mentioned earlier, multiplication of series is somehow troublesome and do not behave in the same way as finite sums (or even as sequences). In this paragraph we discuss this difference and understand under what conditions do products of series converge. We first define the *Cauchy product* of two series:

Definition 3.12.1. Let $\sum a_n$ and $\sum b_n$ be two series. We call the series $\sum c_n$, where

$$c_n = a_0 b_n + a_1 b_{n-1} + \cdots + a_n b_0 = \sum_{j=0}^{n} a_j b_{n-j},$$

the *Cauchy product* of $\sum a_n$ and $\sum b_n$.

A natural question that arises from Definition 3.12.1 is under what conditions is the Cauchy product of two series a convergent series? The following theorem helps us obtain some control on the convergence of the $\sum c_n$.

Theorem 3.12.2 (Mertens' theorem). *Let*

$$\sum_{n=0}^{\infty} a_n = a \in \mathbb{R} \text{ and } \sum_{n=0}^{\infty} b_n = b \in \mathbb{R}$$

be two convergent series. If moreover one of the series (at least) converges absolutely, then the Cauchy product of $\sum a_n$ and $\sum b_n$converges to $a \cdot b$.

The proof of Theorem 3.12.2 is ommited since it is technical and goes beyond the scope of these lecture notes. We emphasize on the fact that at least one series out of $\sum a_n, \sum b_n$ should be absolutely convergent so that the Cauchy product converges too. The following example stresses out this fact.

Example 3.12.3. Consider the series $a_n = b_n = (-1)^n \frac{1}{\sqrt{n+1}}$. These series converge conditionally, since by Leibniz's test we have that the series converge, but by the Integral test (p-series) we have that $\sum \frac{1}{\sqrt{n+1}}$ diverges. Consider now the Cauchy product of these two series:

$$c_n = (-1)^n \sum_{j=0}^{n} \frac{1}{\sqrt{(j+1)(n-j+1)}} \geq \sum_{j=0}^{n} \frac{1}{\sqrt{(n+1)(n+1)}} = 1.$$

We observe that the Cauchy product $\sum c_n$ diverges from the Test for Divergence.

We have proved the following result, which highlights the importance of Theorem 3.12.2.

If both series are only conditionally convergent, the Cauchy product series need not even converge.

Exercise 3.12.4. Let $\sum_{n=0}^{\infty} a^n$ be a series such that $0 < a < 1$. Prove that $\left(\sum_{n=0}^{\infty} a^n\right)^2$ converges.

Review Exercises on Series

1. Prove that the series $\sum_{n=0}^{\infty} \frac{1}{n!}$ converges.

2. Find a series that converges but does not converge absolutely.

3. Determine whether each of the following series is convergent or divergent:

$$i. \quad \sum_{n=1}^{\infty} \frac{\sin n\theta}{n^2} \qquad ii. \quad \sum_{n=1}^{\infty} \frac{\cos n\pi}{\sqrt{n}}$$

$$iii. \quad \sum_{n=1}^{\infty} (-1)^n \frac{\log n}{n} \qquad iv. \quad \sum_{n=1}^{\infty} \frac{n^2}{n!}$$

$$v. \quad \sum_{n=1}^{\infty} \frac{n^4}{4^n} \qquad vi. \quad \sum_{n=2}^{\infty} \frac{1}{n \ln n}$$

$$vii. \quad \sum_{n=2}^{\infty} \frac{(-1)^n}{\ln n} \qquad viii. \quad \sum_{n=1}^{\infty} \frac{n!}{n^n}$$

$$ix. \quad \sum_{n=1}^{\infty} \frac{2^n n!}{n^n} \qquad x. \quad \sum_{n=1}^{\infty} \frac{2}{\sqrt{n(n+1)}}$$

$$xi. \quad \sum_{n=1}^{\infty} \frac{n}{n^3 - 4} \qquad xii. \quad \sum_{n=1}^{\infty} \frac{n!}{(2^n)^3}$$

$$xiii. \quad \sum_{n=1}^{\infty} \frac{1}{n^{1+1/n}} \qquad xiv. \quad \sum_{n=1}^{\infty} \frac{(-1)^n n}{n+1}$$

$$xv. \quad \sum_{n=0}^{\infty} \frac{(-2)^n}{5^{n+1}}$$

Review Test on Series

1. State whether the following are true or false. If false, then provide a counter example.

   i. If $a_n \to 0$, then $\sum_{n=1}^{\infty} a_n$ converges.

   ii. If $\sum_{n=1}^{\infty} |a_n|$ diverges, then $\sum_{n=1}^{\infty} a_n$ also diverges.

   iii. The series $\sum_{n=1}^{\infty} \frac{1}{n^p}$ converges for $p < 1$.

   iv. Assume that a_n and b_n are sequences such that $a_n < b_n$ for all $n \in \mathbb{N}$. If $\sum_{n=1}^{\infty} b_n$ converges, then $\sum_{n=1}^{\infty} a_n$ also converges.

   v. If $\sum_{n=1}^{\infty} a_n$ converges and $\sum_{n=1}^{\infty} b_n$ diverges, then $\sum_{n=1}^{\infty} a_n + b_n$ always diverges.

   vi. If $\sum_{n=1}^{\infty} a_n$ diverges, then $\sum_{n=1}^{\infty} \lambda \cdot a_n$ diverges for all $\lambda \neq 0$.

   vii. If $a_n \geq 0, \forall n \in \mathbb{N}$ and s_n is bounded above, then $\sum_{n=1}^{\infty} a_n$ converges.

   viii. $\sum_{n=1}^{\infty} (-1)^n a_n$ converges when $a_n \to 0$.

   ix. If $\frac{|a_{n+1}|}{|a_n|} < 1$, then $\sum_{n=1}^{\infty} a_n$ converges.

2. By definition, $\sum_{n=1}^{\infty} a_n$ is said to be conditionally convergent if:

3. Let $a_n > 0$ for all $n \in \mathbb{N}$. Prove that $\sum_{n=1}^{\infty} a_n$ converges *if and only if* $\sum_{n=1}^{\infty} \frac{a_n}{1+a_n}$ converges.

4. Let $\sum_{n=1}^{\infty} a_n$ be a convergent series and let $b_n = \sum_{n=m}^{\infty} a_n$, for all $m \in \mathbb{N}$. Prove that $b_n \to 0$.

5. Let $a_n > 0, \forall n \in \mathbb{N}$ and such that $\sum_{n=1}^{\infty} a_n$ converges.

   i. Prove that $\sum_{n=1}^{\infty} a_n^p$ converges for all $p \geq 1$.

   ii. Prove that $\sum_{n=1}^{\infty} a_{k_n}$ converges, where a_{k_n} is a sub-sequence of a_n. (*Hint:* Use the inequality $\sum_{j=1}^{n} a_{k_j} \leq \sum_{j=1}^{k_n} a_j$).

6. Determine whether the following series converge or diverge.

   i. $\sum\limits_{n=0}^{\infty} \frac{n! \cdot (n+1)!}{(2n)!+1}$

   ii. $\sum\limits_{n=0}^{\infty} \left(\frac{n}{n+1}\right)^{n^2}$

   iii. $\sum\limits_{n=2}^{\infty} \left(\frac{1}{n \ln^5 n}\right)$

   iv. $\sum\limits_{n=1}^{\infty} \left(\frac{\sin n}{n^2}\right)$

   v. $\sum\limits_{n=0}^{\infty} \left(\frac{n^2+n+1}{n^4+2n^3-5}\right)$

   vi. $\sum\limits_{n=1}^{\infty} (-1)^n \left(\frac{\ln n}{n}\right)$

Chapter 4

Pointwise and uniform convergence

Sequences are considered so far as *ordered list of numbers*. However, a very useful concept in analysis is to use sequences of functions. In this chapter we describe two different notions of a *limit* of such sequences.

4.1 Pointwise Convergence

Definition 4.1.1. For every $n \in \mathbb{N}$ let $f_n \colon S \to \mathbb{R}$ be a function. We say the sequence $\{f_n\}_{n=1}^{\infty}$ *converges point-wise* to $f \colon S \to \mathbb{R}$, if $\forall x \in S$ we have that:

$$f(x) = \lim_{n\to\infty} f_n(x).$$

Remark 4.1.2. As we have seen earlier, the limit of a sequence, if exists, it is unique. Thus, the function f is unique.

Example 4.1.3. Consider the sequence of functions $f_n(x) = x^{2n+1}$. On the interval $[-1, 1]$, this sequence converges point-wise to

$$f(x) = \begin{cases} 1 & \text{if } x = 1, \\ -1 & \text{if } x = -1, \\ 0 & \text{otherwise.} \end{cases}$$

This is obvious, since for $x = 1$ we have that $f_n(1) = 1 \Rightarrow f_n(x) \to 1$, for $x = -1$ we have that $f_n(-1) = -1 \Rightarrow f_n(x) \to -1$ and for all $x \in (-1, 1)$ we have that $f_n(x) \to 0$.

As we will soon see, functions are usually presented as series and we will use the the notion of point-wise convergence to find the values of the function.

Example 4.1.4. Another example worth presenting here is the following:

$$f_n(x) := \cos(nx).$$

We can see that f_n does not converge point-wise to any function on *any interval*. The function converges at certain points: $x = \frac{\pi}{2} + 2k\pi$ for all $k \in \mathbb{Z}$, *but* not in any interval $[a, b]$, since $\cos(nx)$, when $\cos(nx) \neq 0$, diverges (and in particular the limit doesn't exist, since it oscillates from -1 to 1).

Using now the definition of convergence of a sequence of real numbers, we have the following *equivalent* definition for point-wise convergence:

Definition 4.1.5. Let $f_n\colon S \to \mathbb{R}$ and $f\colon S \to \mathbb{R}$ be functions. Then $\{f_n\}$ converges point-wise to f if and only if

$$\forall\, x \in S \;\&\; \forall\, \epsilon > 0,\; \exists\, n_0 \in \mathbb{N} \;:\; |f_n(x) - f(x)| < \epsilon\, \forall\, n \geq n_0.$$

Remark 4.1.6. We emphasize on the fact that the $n_0 \in \mathbb{N}$ of Definition 4.1.5 depends on ϵ **and** x.

In the next paragraph we discuss the situation where n_0 depends **only** on ϵ. This leads to the notion of *uniform convergence.*

4.2 Uniform Convergence

Definition 4.2.1. Let $f_n\colon S \to \mathbb{R}$ and $f\colon S \to \mathbb{R}$ be functions. We say the sequence $\{f_n\}$ *converges uniformly* to f, if:

$$\forall\, \epsilon > 0\, \exists\, n_0 \in \mathbb{N} \;:\; |f_n(x) - f(x)| < \epsilon,\; \forall\, x \in S.$$

It is obvious from the definition that uniform convergence is understood as convergence where the existence of $n_0 \in \mathbb{N}$ depends only on a given $\epsilon > 0$ and not on x. Thus, uniform convergence implies pointwise convergence. That is:

Let $\{f_n\}$ be a sequence of functions $f_n\colon S \to \mathbb{R}$. If $\{f_n\}$ converges uniformly to $f\colon S \to \mathbb{R}$, then $\{f_n\}$ converges point-wise to f.

The following example shows that the converse of this result is not true in general.

Example 4.2.2. In Example 4.1.3 we considered the functions $f_n(x) = x^{2n+1}$ and proved that these functions converge pointwise on $[-1, 1]$. We show now that they don not converge uniformly on that interval. Indeed, assume that they converge uniformly. Then, since for all

$$\epsilon > 0,\; \exists\, n_0 \in \mathbb{N} \;:\; |f_n(x) - f(x)| < \epsilon,\; \forall\, x \in [-1, 1],$$

we can find such n_0 for $\epsilon = 1/2$. Then $x^{2n_0+1} = |x^{2n_0+1} - 0| < 1/2$ for all $x \in (-1, 1)$, since $f_n(x) \to 0$, $\forall x \in (-1, 1)$. But that means that for any sequence $\{x_k\}$ in $(-1, 1)$ such that $\lim x_k = 1$ we have $x_k^{2n_0+1} < 1/2$ for all k. On the other hand x^{2n_0+1} is a continuous function of x (it is a polynomial), therefore we obtain a contradiction

$$1 = 1^{2n_0+1} = \lim_{k\to\infty} x_k^{2n_0+1} \leq 1/2.$$

However, if we restrict our domain to $[-a, a]$ where $0 < a < 1$, then $\{f_n\}$ converges uniformly to 0 on $[-a, a]$. First note that $a^{2n} \to 0$ as $n \to \infty$. Thus given $\epsilon > 0$, pick $n_0 \in \mathbb{N}$ such that $a^{2n+1} < \epsilon$ for all $n \geq n_0$. Then for any $x \in [-a, a]$ we have $|x| \leq a$. Therefore, for $n \geq n_0$

$$|x^{2n_0+1}| = |x|^{2n_0+1} \leq a^{2n_0+1} < \epsilon.$$

We conclude this chapter by defining *uniformly Cauchy sequences*:

Definition 4.2.3. Let $f_n : S \to \mathbb{R}$ be bounded functions. The sequence of functions f_n is *uniformly Cauchy* if

$$\forall\, \epsilon > 0,\ \exists\, n_0 \in \mathbb{N}\ :\ |f_m - f_k| < \epsilon,\ \forall\, m, k \geq n_0.$$

Exercises 4.2.4. Convergence

1. Prove that if a sequence of functions $f_n \colon S \to \mathbb{R}$ converges uniformly to a bounded function $f \colon S \to \mathbb{R}$, then $\exists\, n_0 \in \mathbb{N}$ such that $\forall n \geq n_0$, f_n are bounded.

2. Let $\{f_n(x)\}$ be a sequence of continuous functions on $S \subset dom(f)$. If $f_n(x)$ converges uniformly to $f(x)$ on S, prove that $f(x)$ is continuous on S.

3. For each $n \in \mathbb{N}$, let $f_n \colon [0, 1] \to \mathbb{R}$ be given by $f_n(x) = x^n$. Determine what function the sequence (f_n) converges to pointwise. Does this sequence converge uniformly to this function?

4. Let $f \colon A \to \mathbb{R}$ be a function. We say that f is *uniformly continuous* if $\forall\, \epsilon > 0$, $\exists\, \delta > 0$ such that for all $x, y \in A$

   $$\text{if } |x - y| < \delta, \quad \text{then } |f(x) - f(y)| < \epsilon.$$

   Determine whether the following functions f are uniformly continuous on the given intervals A:

   i. $f(x) = x^2$, $A = \mathbb{R}$
   ii. $f(x) = x^2$, $A = (-1, 1)$
   iii. $f(x) = \sqrt{x}$, $A = [0, +\infty)$
   iv. $f(x) = \sqrt{x}$, $A = [1, +\infty)$

Chapter 5

Power Series

In this Chapter we express known functions as infinite sums that *resemble* polynomials. This technique is quit useful especially for simplifying expressions, for integrating *non-elementary* functions, for solving *differential equations* and more. We start from the Taylor polynomials and the generalize those polynomials to Taylor Series. Then, we study more general series called Power Series.

5.1 Taylor Polynomials & Series

A classical technique of calculus (and most other fields of mathematics) is to use "simple" functions to obtain local information about relatively complicated functions. The idea is based on the *linearization* of a function (first derivative), where its tangent line is used in order to approximate values of the function. Taylor refined this idea and used polynomials of arbitrary degree in place of the first degree polynomials describing tangent lines. Brook Taylor was a famous musician, painter and also a mathematician. In this subsection we will present and construct the *Taylor & Maclaurin* polynomials for functions. Polynomials comprise a special family of functions since that they are infinitely many times differentiable and integrable. In that sense, the use of polynomials to approximate arbitrary functions becomes essential in Mathematical Analysis.

5.1.1 Motivation & Taylor polynomials

Let f be a function on an interval $[a, b]$ where we assume now that it is $(n+1)$-times differentiable. Our aim is to find a polynomial $p(x)$ that not only approximates $f(x)$ at a, but also the first n derivatives of f at a. That is:

$$f(a) = p(a), \quad f'(a) = p'(a), \quad \ldots, \quad f^{(n)}(a) = p^{(n)}(a),$$

where $f^{(n)}$ represents the nth derivative of f. We start with the *linearization* of f at a, since we have that $f(a) = p(a)$ (the tangent line at a of the graph of f). We denote this polynomial by p_1 since we only use the first derivative of f and it is a polynomial of *degree* one. We have that:

$$p_1(x) = f(a) + f'(a)(x-a), \quad p_1(a) = f(a), \quad p_1'(a) = f'(a)$$

If f has a lot of *curvature* near a, then p_1, the tangent line of C_f at a, may not provide a good approximation. For that reason we consider a *quadratic approximation* of f by simply adding a term of the form $c \cdot (x-a)^2$ and set

$$p_2(x) = f(a) + f'(a)(x-a) + c \cdot (x-a)^2.$$

To determine c we work as follows:

$$p_2(a) = f(a), \quad p_2'(a) = f'(a), \quad p_2''(a) = 2c = f''(a) \Rightarrow c = \frac{1}{2} \cdot f''(a)$$

So, $p_2(x) = f(a) + f'(a)(x-a) + \frac{1}{2} \cdot f''(a)(x-a)^2$. We now extend this process and we have that $p_3(x) = p_2(x) + d(x-a)^3$ and for the same reasons, $d = \frac{1}{6} \cdot f'''(a)$. Continuing in that way obtain the nth- approximating polynomial for f at a, which is called the *Taylor polynomials*. It is worth mentioning here that the same pattern occurs no matter what the original function is. This leads to the following definition:

Definition 5.1.1. Let f be a function such that $f', f'', \ldots, f^{(n)}$ are defined at a. The nth-order *Taylor polynomial* of f centered at a and denoted as p_n, is:

$$p_n(x) = f(a) + \frac{f'(a)}{1!} \cdot (x-a) + \frac{f''(a)}{2!} \cdot (x-a)^2 + \ldots + \frac{f^{(n)}(a)}{n!} \cdot (x-a)^n \quad (5.1)$$

and has the following properties:

$$p_n(a) = f(a),\ p_n'(a) = f'(a),\ p_n''(a) = f''(a),\ ,\quad p_n^{(n)}(a) = f^{(n)}(a)$$

In other words, $p_n(x) = \sum\limits_{k=0}^{n} \frac{f^{(k)}(a)}{k!} \cdot (x-a)^k$.

If $a = 0$, then the polynomial is called the nth-order *Maclaurin* polynomial of f.

Example 5.1.2. We will prove that for $f(x) = e^x$, the Taylor polynomials centered at zero are given by the formula $p_n(x) = \sum\limits_{k=0}^{n} \frac{x^k}{k!}$. Since f is continuous and differentiable on $\mathbb{R}$ with $f^{(n)}(x) = e^x, \forall\, n \in \mathbb{N}$, we have that $f^{(n)}(0) = 1$ and from Equation 5.1 we get:

$$p_n(x) = f(0) + \frac{f'(0)}{1!} \cdot x + \frac{f''(0)}{2!} \cdot x^2 + \ldots + \frac{f^{(n)}(0)}{n!} \cdot x^n = \sum_{k=0}^{n} \frac{x^k}{k!}$$

As in the case of all kinds of approximations, when using the nth-order Taylor polynomial to approximate a function f, an "error" is involved. We now turn our attention to understanding how "good" this approximation is. Denote the *error* made in approximating f by p_n at the point x, by $|R_n(x)|$. Then, we call the difference $f(x) - p_n(x)$ the *remainder*. We have the following result:

Theorem 5.1.3 (Lagrange's Formula for the Remainder Term). *Let f be a functions that has continuous derivatives up to $f^{(n+1)}$ on an open interval I containing a. For all $x \in I$, $f(x) = p_n(x) + R_n(x)$, where p_n is the nth-order Taylor polynomial for f centered at a. The remainder is given by the formula:*

$$R_n(x) = \frac{f^{(n+1)}(c)}{(n+1)!} \cdot (x-a)^{n+1}$$

for some point c between x and a.

The theoretical background about Taylor polynomials is the Mean Value Theorem. It asserts that, *for a function f continuous on an interval $[a, b]$ and differentiable on (a, b), there is a point $c \in (a, b)$ such that*

$$f'(c) = \frac{f(b) - f(a)}{b - a}.$$

The proof of Theorem 5.1.3 is left as an exercise.

Corollary 5.1.4 (Taylor; Estimate of the Remainder). *If $|f^{(n+1)}(x)| \leq M$ for $|x - a| \leq m$, then the remainder $R_n(x)$ of the Taylor polynomial satisfies the following inequality:*

$$|R_n(x)| \leq \frac{M}{(n+1)!} |x - a|^{n+1}$$

Proof. Assume that $f''(x) \leq M$. Then we have that:

$$\int_a^x f''(x)\, dx \leq M(x - a) \Rightarrow f'(x) - f'(a) \leq M(x - a) \Rightarrow$$

$$f'(x) \leq f'(a) + M(x - a) \Rightarrow$$

$$\int_a^x f'(t)\, dt \leq \int_a^x (f'(a) + M(t - a))\, dt \Rightarrow$$

$$f(x) - f(a) - f'(a)(x - a) \leq \tfrac{M}{2} (x - a)^2 \Rightarrow$$

$$R_1(x) \leq \tfrac{M}{2} (x - a)^2$$

Similarly, assuming that $f''(x) \geq -M$, we get that

$$R_1(x) \geq \frac{M}{2} (x - a)^2, \ i.e.\ R_1(x) = \frac{M}{2} (x - a)^2.$$

Apply the same technique $(n + 1)$-times and the result follows. □

An alternative formula of the remainder is given in the following theorem:

Theorem 5.1.5 (Integral form for the Remainder Term). *If $f^{(n+1)}$ is continuous on an interval I and $x \in I$, then:*

$$R_n(x) = \int_0^x \frac{(x - t)^n}{n!} f^{(n+1)}(t)\, dt$$

Proof. The proof is based on the integration by parts technique of evaluating integrals. We sketch the steps of the proof and the reader is adviced to fill in the details. We have that:

$$\begin{aligned}
f(x) &= f(0) + \int_0^x f'(t)dt \\
&= f(0) + \int_0^x -(x - t)' f'(t)dt \\
&= f(0) + \int_0^x (x - t)''(t)dt - [(x - t)f'(t)]_0^x \\
&= f(0) + \int_0^x (x - t)f''(t)dt - f'(0)x \\
&= f(0) + f'(0)x + \int_0^x (x - t)f''(t)dt = \ldots
\end{aligned}$$

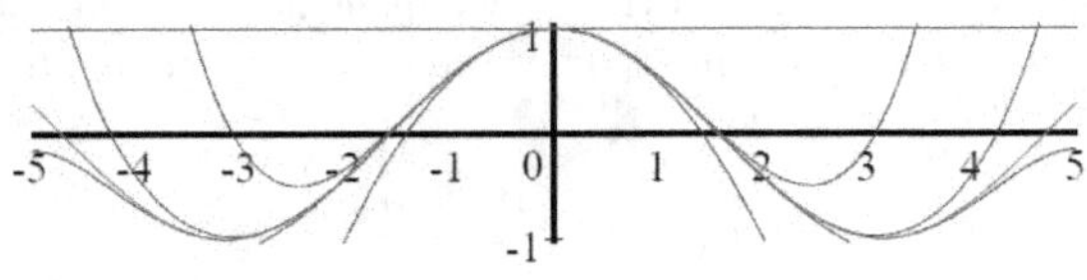

Figure 5.1: The Taylor polynomials for $f(x) = \cos x$

.

□

Exercises 5.1.6. Taylor & Maclaurin Polynomials

1. Find the nth degree Taylor polynomial for the following functions:

   i. $f(x) = \ln x$ centered at $x = 1$.

   ii. $f(x) = e^x$ centered at $x = 0$.

   iii. $f(x) = \ln(1 + x)$ centered at $x = 0$.

   iv. $f(x) = \sin x$ centered at $x = 0$.

2. Use the third order Maclaurin polynomial for the function $f(x) = \sin x$ to approximate $\sin 1$ and determine the accuracy of this approximation.

3. What Maclaurin polynomial could be used in order to approximate $\sqrt{e}$ within 0.01?

4. Find the Taylor polynomials for:

   i. $\left(\frac{x^2}{(1-x)^2}\right)$

   ii. $\left(\log \frac{(1-x)^3}{(1+x)^5}\right)$

   iii. $(e^x \cos x)$

5. Evaluate the Maclaurin polynomial for $f(x) = \cos x$. See Figure 5.1

5.1.2 Taylor & Maclaurin Series

In this subsection we will see how for many functions $f(x)$, the Taylor polynomials $p_n(x)$ give better and better approximations as you add more terms. For this reason the limit when $n \to \infty$ is considered, which leads to the *infinite sum*:

$$f(x) = f(a) + f'(a) \cdot (x - a) + \frac{f''(a)}{2!} \cdot (x - a)^2 + \frac{f'''(0)}{3!} \cdot (x - a)^3 + \dots$$

We can think of it as a series of functions where the nth partial sum of the series coincides with the nth order Taylor polynomial of the function f around a real number a. In that sense, the Taylor series of a function f encodes all of its Taylor polynomials around a.

Consider for example the function $f(x) = \sin x$. This function is infinitely many times differentiable and its derivatives satisfy:

$$f(x) = \sin x, \quad f'(x) = \cos x, \quad f''(x) = -\sin x, \quad f^{(3)}(x) = -\cos x,$$

and thus

$$f^{(4)}(x) = \sin x = f(x).$$

We observe the following pattern:

$$\sin x, \ \cos x, \ -\sin x, \ -\cos x, \ \sin x, \ \cos x, \ -\sin x, \ -\cos x, \ \sin x, \ \ldots$$

Let us consider now the values of $f^{(i)}(x)$ at $x = 0$, $f^{(j)}(0)$. We have:

i	1	2	3	4	5	6	7	8	$\cdots$
$f^{(i)}(0)$	0	1	0	-1	0	1	0	-1	$\cdots$

Thus, we obtain the following Taylor polynomials:

$$\begin{aligned} p_0(x) &= 0 \\ p_1(x) &= x \\ p_2(x) &= x \\ p_3(x) &= x - \frac{x^3}{3!} \\ p_4(x) &= x - \frac{x^3}{3!} \\ p_5(x) &= x - \frac{x^3}{3!} + \frac{x^5}{5!} \end{aligned}$$

It is worth mentioning that since $f^{(2)}(0) = 0 = f^{(4)}(0)$,

$$p_1(x) = p_2(x), p_3(x) = p_4(x) \ etc.$$

We now have that:

$$p_n(x) = \sum_{i=1}^{n} (-1)^n \frac{x^{2n+1}}{(2n+1)!} \simeq \sin x$$

and we observe that $f^{(n+1)}(x)$ is either $\pm \sin x$ or $\pm \cos x$. Thus,

$$|f^{(n+1)}(x)| \leq 1, \forall x.$$

Considering $M = 1$ in the remainder formula we have that

$$|R_n(x)| \leq \frac{|x|^{n+1}}{(n+1)!}.$$

We will prove that $R_n \to 0$. Indeed, consider the series $\sum \frac{|a|^{n+1}}{(n+1)!}$, for $a \in \mathbb{R}$. Applying the Ratio Test we get that:

$$\frac{\frac{|a|^{n+2}}{(n+2)!}}{\frac{|a|^{n+1}}{(n+1)!}} = \frac{|a|}{(n+2)} \to 0 < 1 \Rightarrow \sum \frac{|a|^{n+1}}{(n+1)!} \text{ converges} \Rightarrow$$

$$\frac{|a|^{n+1}}{(n+1)!} \to 0, \forall\, a \in \mathbb{R}$$

Thus, $|R_n(x)| \leq \frac{|x|^{n+1}}{(n+1)!} \to 0$, i.e., the remainder tends to zero as n becomes arbitrarily large. It is clear that this means that for fixed $x_0 \in \mathbb{R}$, the series $\sum_{i=1}^{n}(-1)^n \frac{x_0^{2n+1}}{(2n+1)!}$ converges to $\sin x_0$, and this is true for all x_0. We then say that the series $\sum_{i=1}^{n}(-1)^n \frac{x^{2n+1}}{(2n+1)!}$ converges to the function $\sin x$, denoted by

$$\sum_{i=1}^{n}(-1)^n \frac{x^{2n+1}}{(2n+1)!} = \sin x.$$

Example 5.1.7. We now evaluate the Taylor series of $f(x) = \ln x$ around $x = 1$. We have that:

$$\begin{aligned}
f(x)^{(0)} &= \ln x, & f^{(0)}(1) = 0 \\
f(x)' &= x^{-1}, & f'(1) = 1 \\
f(x)'' &= -x^{-2}, & f''(1) = -1 \\
f(x)''' &= 2x^{-3}, & f'''(1) = 2 \\
f(x)^{(4)} &= -6x^{-4}, & f''''(1) = -6
\end{aligned}$$

In general, the nth derivative equals $\pm(n-1)!$, so the coefficients of the Taylor series are $\pm(n-1)!/n! = \pm 1/n$, except for the $n = 0$ term, which is zero. The resulting series is

$$\ln x = (x-1) - \frac{1}{2}(x-1)^2 + \frac{1}{3}(x-1)^3 + \ldots$$

It is left as an exercise to prove that the power series above actually converges to $\ln x$.

Remark 5.1.8. Given a function f we may be able to compute the Taylor series, but that does not mean we have found a series representation for f. We need to know where the series converges and if it converges to $f(x)$.

Below are some common Taylor series:

$$\begin{aligned}
\frac{1}{1-x} &= \sum_{n=0}^{\infty} x^n = 1 + x + x^2 + x^3 + \cdots \quad \text{(when } -1 < x < 1\text{)} \\
e^x &= \sum_{n=0}^{\infty} \frac{x^n}{n!} = 1 + x + \frac{x^2}{2!} + \frac{x^3}{3!} + \cdots \\
\sin x &= \sum_{n=0}^{\infty} \frac{(-1)^n}{(2n+1)!} x^{2n+1} = x - \frac{x^3}{3!} + \frac{x^5}{5!} - \cdots \\
\cos x &= \sum_{n=0}^{\infty} \frac{(-1)^n}{(2n)!} x^{2n} = 1 - \frac{x^2}{2!} + \frac{x^4}{4!} - \cdots
\end{aligned}$$

Exercises 5.1.9. Taylor Series

1. Expand $(1+x)^{1/3}$ in a Taylor series around $x=0$.

2. Find the Taylor series expansion of $\log_2 x$ around $x=1$.

3. For each function, find the Maclaurin series or Taylor series centered at a, and the radius of convergence.

$i.\ f(x)=\cos x$ $\qquad ii.\ f(x)=1/x,\ x_0=5$

$iii.\ f(x)=\ln x,\ x_0=2$ $\qquad iv.\ f(x)=1/x^2,\ x_0=1$

$v.\ f(x)=\sqrt{1-x},\ x_0=0$ $\qquad vi.\ f(x)=\tan x,\ x_0=0$

4. Find the Taylor series for the following functions and determine its radius of convergence:

(i) $f(x)=\log(1+x)$

(ii) $f(x)=\dfrac{1}{1+x}$

(iii) $f(x)=\arctan x$

5.2 Power Series

5.2.1 Motivation & Definition

Consider the series

$$\sum_{n=0}^{\infty}\frac{2^n}{n!},\quad \sum_{n=0}^{\infty}\frac{3^n}{n!},\quad \sum_{n=0}^{\infty}\frac{e^n}{n!},\quad \sum_{n=0}^{\infty}\frac{\pi^n}{n!},\quad \sum_{n=0}^{\infty}\frac{(-1)^n}{n!} \tag{5.2}$$

All these series look very similar, since only the number on the power changes. If we wish to determine whether these series converge or diverge, we need to apply the Ratio Test on each one of them and decide. But it is natural to wonder whether we can deal with all of these series at once or not. Towards an answer to that question, consider the series

$$\sum_{n=0}^{\infty}\frac{x^n}{n!},\qquad \text{where } x\in\mathbb{R} \text{ is a constant number.}$$

We will use the Ratio Test for this series. We set up the appropriate limit first and since a_n is not necessarily positive, we will consider absolute values for the quotient $\frac{a_{n+1}}{a_n}$, where $a_n=\frac{x^n}{n!}$. We have that:

$$\lim_{n\to\infty}\left|\frac{\frac{x^{n+1}}{(n+1)!}}{\frac{x^n}{n!}}\right| = \lim_{n\to\infty}\left|\frac{x^n\cdot x\cdot n!}{x^n\cdot(n+1)\cdot n!}\right| = \lim_{n\to\infty}\frac{|x|}{n+1} = 0.$$

Since $\left|\frac{a_{n+1}}{a_n}\right| \to 0 < 1$, we get that the series $\sum\limits_{n=1}^{\infty} \frac{x^n}{n!}$ converges for all $x \in \mathbb{R}$. So, all the series in Equation 5.2 converge. We now observe that for each fixed x, the series $\sum\limits_{n=0}^{\infty} \frac{x^n}{n!}$ converges to a real number $l \in \mathbb{R}$ and so we get a correspondence between two subsets of $\mathbb{R}$, i.e. we obtained a *function*. The series $\sum\limits_{n=0}^{\infty} \frac{x^n}{n!}$ is an example of a *power series*.

$$\sum_{n=0}^{\infty} \frac{x^n}{n!} = 1 + x + \frac{x^2}{2!} + \frac{x^3}{3!} + \frac{x^4}{4!} + \dots$$

The reason why we call these kind of series, power series, is because they are sums of powers of x, and thus, they *resemble polynomials*. They are not polynomials since they don't have a *degree* (or their is degree is infinite) and as we will soon see, their domain is not necessarily $\mathbb{R}$. In general we can say that a power series is much more complicated than a polynomial.
The questions that rise now are the following:

- *What is the domain of a power series?*
- *Do power series and polynomials have the same kind of properties?*

Before we answer these questions, we will first state the definition of a power series.

Definition 5.2.1. A *power series* is a series of the form

$$\sum_{n=0}^{\infty} a_n \cdot (x-c)^n = a_0 + a_1 \cdot (x-c) + a_2 \cdot (x-c)^2 + \dots$$

where x is a **variable** and the a_n's are constants called the **coefficients** of the series. The number c is called the **center** of the power series.

It is worth mentioning that for each fixed x, a power series becomes a series of constants that we can test for convergence or divergence using the tests we studied in the previous chapter.

5.2.2 Domain of a Power Series

It is natural to consider the domain of a power series to consist of those x's for which the series of constants is convergent.

Example 5.2.2. i. We will check for convergence the power series

$$\sum_{n=0}^{\infty} (n^2+1) \cdot x^n = 1 + 2x + 5x^2 + \dots.$$

As noted before, we will set up the appropriate limit in order to use the Ratio Test. We have that:

$$\lim_{n\to\infty}\left|\frac{a_{n+1}}{a_n}\right| = \lim_{n\to\infty}\left|\frac{(n^2+2n+2)\cdot x^{n+1}}{(n^2+1)\cdot x^n}\right| = |x|.$$

The ratio test says that the series converges provided that this limit is less than 1, that is, when $|x| < 1$. So, the power series converges when $-1 < x < 1$, diverges when $x > 1$ or $x < -1$, and we can not conclude for $x = 1$ or $x = -1$. We now check the endpoints 1 and -1 separately:

For $x = 1$ we have that the series $\sum\limits_{n=0}^{\infty}(n^2+1)$ diverges from the Test for Divergence, since $a_n = n^2 + 1 \to +\infty$.

For $x = -1$ we have: $\sum\limits_{n=0}^{\infty}(n^2+1)\cdot(-1)^n$. This series diverges again from the Test for Divergence, since for the subsequences we have:

$$a_{2n} = (4n^2+1) \to +\infty \text{ and } a_{2n+1} = -(4n^2+4n+1) \to -\infty.$$

We conclude that the domain of the power series $\sum\limits_{n=0}^{\infty}(n^2+1)\cdot x^n$ is the open interval $(-1, 1)$.

ii. We will check for convergence the power series $\sum\limits_{n=0}^{\infty}(-1)^n\frac{(x+1)^n}{n!}$. We set up the appropriate limit first:

$$\lim_{n\to\infty}\left|\frac{a_{n+1}}{a_n}\right| = \lim_{n\to\infty}\left|\frac{\frac{(x+1)^{n+1}}{(n+1)!}}{\frac{(x+1)^n}{n!}}\right| = \lim_{n\to\infty}\left|\frac{(x+1)\cdot n!}{(n+1)!}\right| = 0.$$

Since the limit is 0 (which is less than 1), the ratio test says that the series converges absolutely for all x, and thus, the domain of this power series $\mathbb{R}$.

iii. Consider now the power series $\sum\limits_{n=0}^{\infty}(-1)^n\frac{(x+1)^n}{n}$. We set up the appropriate limit and we have:

$$\lim_{n\to\infty}\left|\frac{a_{n+1}}{a_n}\right| = \lim_{n\to\infty}\left|\frac{\frac{(x+1)^{n+1}}{(n+1)}}{\frac{(x+1)^n}{n}}\right| = \lim_{n\to\infty}\left|\frac{(x+1)\cdot n}{(n+1)}\right| = |x+1|.$$

The ratio test says that the series converges provided that this limit is less than 1, that is, when $|x+1| < 1$. So, the power series converges when $-2 < x < 0$, diverges when $x > 0$ or $x < -2$, and we can not conclude for $x = 0$ or $x = -2$. We now check the endpoints -2 and 0 separately:

For $x = -2$ we have that the series $\sum\limits_{n=0}^{\infty}\frac{1}{n}$ diverges (Harmonic Series).

For $x = 0$ we have that the $\sum_{n=0}^{\infty} \frac{(-1)^n}{n}$ converges (Alternating Harmonic Series). We conclude that the domain of this power series is the interval $(-2, 0]$.

iv. Consider now the power series $\sum_{n=0}^{\infty} (n+1)! \cdot x^n$. We set up the appropriate limit first:

$$\lim_{n\to\infty} \left| \frac{a_{n+1}}{a_n} \right| = \lim_{n\to\infty} \left| \frac{(n+2)! \cdot x^{n+1}}{(n+1)! \cdot x^n} \right| = \lim_{n\to\infty} (n+2) \cdot |x| = +\infty.$$

Since the limit is ∞ (which is greater than 1), the ratio test says that the series diverges for all $x \neq 0$, and thus, the domain of this power series $\{0\}$.

From the Examples 5.2.2 we observe that a power series $\sum_{n=0}^{\infty} a_n \cdot (x-c)^n$ may either converge for all x's in an interval of the form $(c-\rho, c+\rho)$, $[c-\rho, c+\rho)$, $(c-\rho, c+\rho]$, $[c-\rho, c+\rho]$, or it may converge for all $x \in \mathbb{R}$, or it may converge only for $x = c$.

Remark 5.2.3. We note that instead of the Ratio Test, one may choose to use the Root Test and all the above conclusions remain the same.

One interesting way of seeing all the above is the following: Let

$$\sum_{n=0}^{\infty} a_n \cdot (x-c)^n.$$

Then:

$$\lim \left| \frac{a_n}{a_{n+1}} \right| = \rho \Rightarrow \begin{cases} dom = \mathbb{R}, & \text{if } \rho = 0 \\ dom = I, & \text{if } \rho \in \mathbb{R}_+^* \\ dom = \{c\}, & \text{if } \rho = \infty \end{cases},$$

where $I = (c-\rho, c+\rho)$, or $[c-\rho, c+\rho)$, $(c-\rho, c+\rho]$ or $[c-\rho, c+\rho]$. Moreover, ρ is called the *radius of convergence* and I is called the *interval of convergence* of the power series.

Remark 5.2.4. The interval of convergence of a power series is the interval that consists of all values of x for which the series converges, i.e. the domain of the power series. Moreover, $\lim \left| \frac{a_n}{a_{n+1}} \right| = \rho$, the radius of convergence of the power series. Finally, we emphasize on the fact that when $\rho \in \mathbb{R}_+^*$, then we have that the power series converges for $x \in (c-\rho, c+\rho)$ and we test the endpoints separately.

The discussion above leads to the following theorem:

Theorem 5.2.5. *For a given power series $\sum_{n=0}^{\infty} a_n \cdot (x-c)^n$ there are exactly three possibilities:*

i. The series converges only for $x = c$, the center of the series, when $\rho = 0$.

ii. The series converges for all $x \in \mathbb{R}$, when $\rho = \infty$.

iii. The series converges when x is in a specific interval, symmetric with respect to c, when $\rho \in \mathbb{R}_+^*$.

Exercises 5.2.6. Radius of Convergence

1. Determine the radius of convergence of the following power series:

   (i) $f(x) = \sum_{n=0}^{\infty} \frac{x^n}{n!}$

   (ii) $f(x) = \sum_{n=0}^{\infty} n!x^n$

   (iii) $f(x) = 1 - x^2 + x^4 - x^6 + x^8 + \cdots = \sum_{n=0}^{\infty} (-1)^n x^{2n}$

2. Evaluate each of the following infinite sums. (Hint: find a function f and a number a such that the series is equal to $f(a)$.)

   (i) $1 - (1/2)^3 + (1/2)^6 - (1/2)^9 + \cdots$

   (ii) $\sum_{n=0}^{\infty} \frac{1}{(2n)!}$

   (iii) $\sum_{n=0}^{\infty} \frac{(-1)^n 2^{2n} \pi^{2n}}{(2n)!}$

   (iv) $\sum_{n=0}^{\infty} \frac{n}{2^n}$

5.2.3 Derivatives and Integrals of Power Series

We have seen that the sum of a power series is a function whose domain is the interval of convergence of the series. It is very important to note that if a power series $\sum_{n=0}^{\infty} a_n \cdot (x - c)^n$ has interval of convergence I, then the function that is defined through the power series is continuous on I. This result is due to the uniformity of convergence (*why?*). Using the same argument, we may always differentiate and integrate a power series **term by term** within its interval of convergence. More precisely we have the following result called term-by-term differentiation and integration:

Theorem 5.2.7. *It is always permissible to differentiate and integrate a power series* $f(x) = \sum_{n=0}^{\infty} a_n \cdot (x - c)^n$ *of radius of convergence* $R > 0$ *and obtain:*

$$f'(x) = \sum_{n=1}^{\infty} a_n \cdot n\,(x - c)^{n-1}$$

$$F(x) = \int f(x)\,dx = c + \sum_{n=0}^{\infty} \frac{a_n}{n+1} \cdot (x - c)^{n+1}$$

The radii of convergence of the power series above are both R.

Remark 5.2.8. Theorem 5.2.7 can be written as follows:

$$\frac{d}{dx}\left(\sum_{n=0}^{\infty} a_n \cdot (x-c)^n\right) = \sum_{n=1}^{\infty} a_n \cdot \frac{d}{dx}(x-c)^n$$

and

$$\int \left(\sum_{n=0}^{\infty} a_n \cdot (x-c)^n\right) dx = c + \sum_{n=0}^{\infty} a_n \cdot \int (x-c)^n \, dx,$$

that is, differentiation (resp. integration) and infinite sums commute. The power series obtained by differentiating a power series with interval of convergence I, has interval of convergence I' and I' may differ from the interval I only at the endpoints, i.e. $I' \subseteq I$ and $\overset{o}{I'} = \overset{o}{I}$, where $\overset{o}{I}$ denotes the *interior* of the set I.

Example 5.2.9. Consider the geometric series:

$$\begin{aligned} \tfrac{1}{1-x} &= \textstyle\sum_{n=0}^{\infty} x^n, \quad |x| < 1 \\ \int \tfrac{1}{1-x}\, dx &= \textstyle\int \left(\sum_{n=0}^{\infty} x^n\right) dx \\ -\ln|1-x| &= \textstyle\sum_{n=0}^{\infty} \frac{1}{n+1} x^{n+1} + c \\ \ln|1-x| &= \textstyle\sum_{n=0}^{\infty} -\frac{1}{n+1} x^{n+1} + c, \quad c \in \mathbb{R} \end{aligned}$$

The series does not converge when $x = 1$ but does converge when $x = -1$ or $1-x = 2$. The interval of convergence is $[-1, 1)$, or $0 < 1 - x \leq 2$, so we can use the series to represent $\ln(x)$ when $0 < x \leq 2$. For example

$$\ln(3/2) = \ln(1-(-1/2)) = \sum_{n=0}^{\infty} (-1)^n \frac{1}{n+1} \frac{1}{2^{n+1}}$$

In the previous example a way of evaluating power series representation of functions from known power series is presented. The next example shows how the operation "composition" also works, in the sense that:

$$\begin{aligned} \tfrac{1}{1-x} &= \textstyle\sum_{n=0}^{\infty} x^n &&, |x| < 1 \\ \tfrac{1}{1-g(x)} &= \textstyle\sum_{n=0}^{\infty} (g(x))^n &&, |g(x)| < 1 \end{aligned}$$

Example 5.2.10. We will find a power series for $\frac{1}{4-4x+x^2}$.

$$\begin{aligned} \tfrac{1}{4-4x+x^2} \overset{fact.}{=} \tfrac{1}{(2-x)^2} &= \left(\tfrac{1}{2-x}\right)' = \\ \overset{sub.}{=} \left(\tfrac{1}{2}\tfrac{1}{1-u}\right)' &= \left(\tfrac{1}{2}\sum_{n=0}^{\infty} u^n\right)' = \\ = \left(\tfrac{1}{2}\sum_{n=0}^{\infty} \left(\tfrac{x}{2}\right)\right)' &= \sum_{n=0}^{\infty} \tfrac{n}{2^{n+1}} x^{n-1} \end{aligned}$$

Remark 5.2.11. Note that not all series functions are differentiable and integrable. Trigonometric series for example (*Fourier series*) are not differentiable. See Figure 5.2.

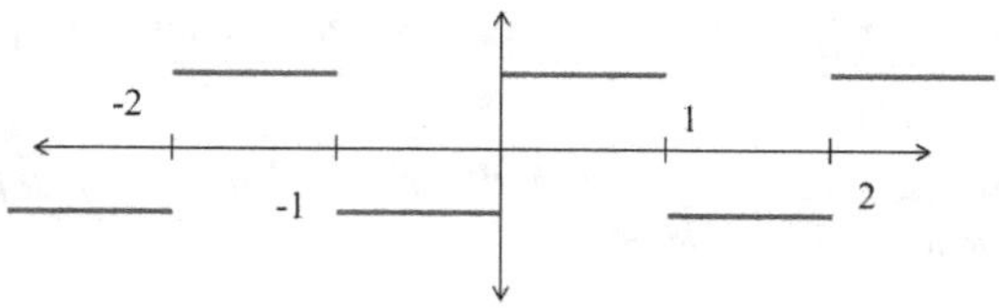

Figure 5.2: An illustration of a Trigonometric (Fourier) series.

Exercises 5.2.12. Differentiating and Integrating Power Series

1. Evaluate a power series presentation for $\arctan x$.
2. Evaluate a power series representation for

$$\int_0^1 \ln(1+x^2)\,dx$$

5.2.4 Multiplying and Dividing Power Series

Sometimes the formula for the nth derivative of a function f is difficult to discover, but a combination of a known Maclaurin series and some algebraic manipulation leads easily to the Maclaurin series for f. In particular, power series behave like polynomials under multiplication and division as illustrated in the following exmple:

Example 5.2.13. We evaluate the Maclaurin series for $x\sin(-x)$.

To get from $\sin x$ to $x\sin(-x)$ we substitute $-x$ for x and then multiply by x. We can do the same thing to the series for $\sin x$:

$$\begin{aligned} x\cdot\textstyle\sum_{n=0}^{\infty}(-1)^n\frac{(-x)^{2n+1}}{(2n+1)!} &= x\cdot\textstyle\sum_{n=0}^{\infty}(-1)^n(-1)^{2n+1}\frac{x^{2n+1}}{(2n+1)!} = \\ &= \textstyle\sum_{n=0}^{\infty}(-1)^{n+1}\frac{x^{2n+2}}{(2n+1)!} \end{aligned}$$

In general, we are only able to find the first few terms of the power series obtained this way (the most important ones), because the calculations for the later terms become tedious.

Example 5.2.14. The power series for $\frac{1}{1-x}\arctan(x)$ can be evaluated by multiplying the power series:

$$\arctan(x) = \sum_{n=0}^{\infty}\frac{(-1)^n x^{2n+1}}{2n+1} \text{ and } \frac{1}{1-x} = \sum_{n=0}^{\infty} x^n$$

$$\begin{aligned} \tfrac{1}{1-x}\arctan(x) &= \left(\textstyle\sum_{n=0}^{\infty} x^m\right)\left(\textstyle\sum_{n=0}^{\infty}\frac{(-1)^n x^{2n+1}}{2n+1}\right) = \\ &= (1+x+x^2+x^3+\ldots)(\tfrac{x}{1}-\tfrac{x^3}{3}+\tfrac{x^5}{5}-\tfrac{x^7}{7}+\ldots) = \\ &= (1)x+(1)x^2+(1-\tfrac{1}{3})x^3+(1-\tfrac{1}{3})x^4+(1-\tfrac{1}{3}+\tfrac{1}{5})x^5+\ldots \end{aligned}$$

We now present the theorem that proves that what we did in the previous examples are legitimate.

Theorem 5.2.15. *Let $f(x) = \sum a_n x^n$ and $g(x) = \sum b_n x^n$ be two power series with radius of convergence ρ. Then the power series*

$$\left(\sum a_n x^n\right) \cdot \left(\sum b_n x^n\right)$$

represents $f(x) \cdot g(x)$ and has interval of convergence ρ.

The proof of Theorem 5.2.15 is omitted.

Remark 5.2.16. Note that Theorem 5.2.15 deals with multiplication of power series only. In order to be able to define $\frac{\sum a_n x^n}{\sum b_n x^n}$ to be a power series presentation of $\frac{f(x)}{g(x)}$, we require $b_0 \neq 0$ and the resulting power series converges for sufficiently small $|x|$.

Exercises 5.2.17. Multiplication and division of Power Series

1. Evaluate a power series presentation for $x \cdot e^{-x}$.
2. Evaluate a power series representation for $\tan x$ using the identity $\tan x = \frac{\sin x}{\cos x}$.

5.3 The Binomial Series

We now focus on functions of the form $f(x) = (1+x)^k$, where $k \in \mathbb{R}$ and we evaluate their Maclaurin series. We have that:

$$\begin{aligned}
f(x) &= (1+x)^k & f(0) &= 1 \\
f'(x) &= k \cdot (1+x)^{k-1} & f'(0) &= k \\
f''(x) &= k \cdot (k-1) \cdot (1+x)^{k-2} & f''(0) &= k \cdot (k-1) \\
f'''(x) &= k(k-1)(k-2)(1+x)^{k-3} & f'''(0) &= k(k-1)(k-2) \\
&\vdots & &\vdots \\
f^{(n)}(x) &= \binom{k}{n}(1+x)^{k-n} & f^{(n)}(0) &= \binom{k}{n}
\end{aligned}$$

where $\binom{k}{n} = \frac{k!}{n! \cdot (k-n)!}$.

Hence, the Maclaurin series for $f(x) = (1+x)^k$ is $\sum_{n=0}^{\infty} \binom{k}{n} x^n$ and it is called the **binomial series**.

We define $\binom{k}{n}$ for negative k in such a way that the identity

$$\binom{k}{n} = \binom{k-1}{n} + \binom{k-1}{n-1}$$

is valid without any restrictions. Then:

Theorem 5.3.1 (Binomial Series). *For $k \in \mathbb{R}$ and $|x| < 1$, we have*

$$(1+x)^k = \sum_{n=0}^{\infty} \binom{k}{n} x^n$$

Remark 5.3.2. It can be shown that the binomial series at 1 if $k \in (-1, 0]$ and it converges at ± 1 if $k \geq 0$.

Exercises 5.3.3. Binomial Series

1. Find the Maclaurin series for the function $f(x) = \frac{1}{\sqrt{1-x}}$.
2. Find the Maclaurin series for the function $f(x) = \frac{x}{(1-x)^{1/3}}$.

5.4 Applications

One very important application that the Taylor series find is in the theory of *non-elementary* functions. Non-elementary functions are functions that although they have anti-derivatives, we are not able to evaluate them in terms of known functions. The most famous example of non-elementary functions is e^{-x^2}. We have that:

$$\begin{aligned}
g(x) &= e^x &&= \sum_{n=0}^{\infty} \frac{x^n}{n!} && \Rightarrow \\
f(x) &= e^{-x^2} &&= \sum_{n=0}^{\infty} \frac{(-1)^n x^{2n}}{n!} && \Rightarrow \\
\int f(x)\,dx &= \int e^{-x^2}\,dx &&= \int \left(\sum_{n=0}^{\infty} \frac{(-1)^n x^{2n}}{n!} \right) dx && = \\
& &&= \sum_{n=0}^{\infty} \frac{(-1)^n x^{2n+1}}{(2n+1)n!} + c
\end{aligned}$$

The series obtained that way converges for all x (*why?*). We can now approximate areas bounded by the graph of f, C_f, the x- axis and the lines $x = a, x = b$, $\int_a^b |f(x)|\,dx$. Consider for example $a = 1$ and $b = 2$. We have that:

$$\int_1^2 f(x)\,dx = \int_1^2 \left(\sum_{n=0}^{\infty} \frac{(-1)^n x^{2n}}{n!} \right) dx = 1 - \frac{1}{3} + \frac{1}{10} - \frac{1}{42} + \ldots$$

Another important use of Taylor series is in the evaluation of limits. Consider for example the limit:

$$\lim_{x \to 0} \frac{e^x - 1}{x}$$

Substituting e^x by its power series presentation, and since power series are continuous functions, we have:

$$\lim_{x\to 0}\frac{e^x-1}{x}=\lim_{x\to 0}\frac{\sum_{n=0}^{\infty}\frac{x^n}{n!}-1}{x}=\lim_{x\to 0}\frac{\sum_{n=1}^{\infty}\frac{x^n}{n!}}{x}=\lim_{x\to 0}\sum_{n=1}^{\infty}\frac{x^{n-1}}{n!}=$$

$$=\lim_{x\to 0}\left(1+\sum_{n=2}^{\infty}\frac{x^{n-1}}{n!}\right)=1$$

Although one could evaluate this limit using other techniques, it is essential to understand the method of power series, that is used in some computer algebra systems.

Power series also find many applications in computer science in order to approximate functions via polynomials, in physics (relativity, electromagnetism, etc), in engineering (buildings, etc), in solving differential equations in general and more.

Review Exercises

1. Determine the radius of convergence and interval of convergence for the following power series:

   i. $\sum_{n=1}^{\infty} \frac{(-1)^n \cdot n}{4^n} \cdot (x+3)^n$

   ii. $\sum_{n=1}^{\infty} \frac{2^n}{n} \cdot (4x-8)^n$

   iii. $\sum_{n=0}^{\infty} n! \cdot (2x+1)^n$

   iv. $\sum_{n=1}^{\infty} \frac{1}{n^n} \cdot (x-6)^n$

   v. $\sum_{n=1}^{\infty} \frac{1}{(-3)^n} (x)^{2n}$

2. Find a power series representation for the following functions and determine their interval of convergence.

   i. $g(x) = \frac{1}{1+x^3}$

   ii. $h(x) = \frac{2x^2}{1+x^3}$

   iii. $k(x) = \frac{x}{5-x}$

   iv. $\sum_{n=1}^{\infty} \frac{1}{n^n} \cdot (x-6)^n$

   v. $\sum_{n=1}^{\infty} \frac{1}{(-3)^n} (x)^{2n}$

   Note that $\sum_{n=0}^{\infty} x^n = \frac{1}{1-x}$, provided $|x| < 1$.

3. Find a power series representation for the following functions and determine their interval of convergence.

   i. $g(x) = \frac{1}{(1-x)^2}$

   ii. $h(x) = \ln(5-x)$

4. Find the Taylor/Maclaurin series for:

   i. $f(x) = e^x$, about $x = 0$.

   ii. $f(x) = e^{-x}$, about $x = 0$.

   iii. $f(x) = x^4 \cdot e^{-3x^2}$, about $x = 0$.

   iv. $f(x) = e^{-x}$, about $x = -4$.

   v. $f(x) = \cos x$, about $x = 0$.

   vi. $f(x) = \sin x$, about $x = 0$.

   vii. $f(x) = \ln x$, about $x = 2$.

viii. $f(x) = \frac{1}{x^2}$, about $x = -1$.

5. Applications of Series:

   i. Determine a Taylor series about $x = 0$ for the function:

   $$\int \frac{\sin x}{x}\, dx.$$

   ii. Find the first 3 non-zero terms in the Taylor series for

   $$f(x) = e^x \cdot \cos x, \text{ about } x = 0.$$

Review Test on Power Series

1. Chose the one choice that best completes the statement or answers the question.

   i. Let $P(x) = 3x^2 - 5x^3 + 7x^4 + 3x^5$ be the fifth-degree Taylor polynomial for the function f about $x = 0$. Then $f'''(0)$ is equal to:

   $a. -30 \quad b. -15 \quad c. -5 \quad d. -\frac{5}{6} \quad e.$ None

   ii. If $\Sigma_{n=0}^{+\infty} a_n \cdot x^n$ converges to $f(x)$ for all $x \in \mathbb{R}$, then $f'(1)$ is equal to:

   $a.\ 0 \quad b.\ a_1 \quad c.\ \Sigma_{n=0}^{+\infty} a_n \quad d.\ \Sigma_{n=1}^{+\infty} n \cdot a_n \quad e.$ None

   iii. The Maclaurin series of $\frac{1}{1-x}$ is $\Sigma_{n=0}^{+\infty} x^n$. Which of the following is a power series expansion for $\frac{x^2}{1-x^2}$?

   $a.\ 1 + x^2 + x^4 + x^6 + \dots$
   $b.\ x^2 + x^3 + x^4 + \dots$ $\qquad c.\ x^2 + 2x^3 + 3x^4 + \dots$
   $d.\ x^2 + x^4 + x^6 + \dots$ $\qquad e.$ None of the above

   iv. Suppose that $\Sigma_{n=0}^{+\infty} a_n \cdot (x-2)^n$ converges for $x = -1$ and diverges for $x = 8$. For which of the following values of x *must* the power series converge?

   $a.\ x = 5\ \&\ x = 3 \quad b.\ x = 2\ \&\ x = 3 \quad c.\ x = 7\ \&\ x = 3$
   $d.\ x = 5\ \&\ x = 2 \qquad e.$ None

   v. Let $\rho = \lim_{n \to +\infty} \frac{|c_n|}{|c_{n+1}|}$. The power series $\sum c_n x^n$ converges when

   $(a)\ x \in [-\rho, \rho] \quad (b)\ x \in (-\rho, \rho)$

   $(c)\ x \in [-\rho, \rho) \quad (d)$ Only when $x = 0$

   vi. Suppose that the power series $\Sigma_{n=0}^{+\infty} c_n \cdot (x-2)^n$ converges when $x = 6$ and diverges when $x = -4$. Then:

   (a) The series $\Sigma_{n=0}^{+\infty} c_n$ diverges.
   (b) The series $\Sigma_{n=0}^{+\infty} (-1)^n \cdot c_n$ converges.
   (c) The series $\Sigma_{n=0}^{+\infty} c_n \cdot 7^n$ converges.
   (d) The series $\Sigma_{n=0}^{+\infty} c_n \cdot 8^n$ converges.
   (e) None of the above.

2. State whether the following statements are true or false.

   i. The radius of convergence remains the same when a power series is integrated or differentiated.

   ii. If f is an infinitely many times differentiable function, then

   $$f(x) = \Sigma_{n=1}^{+\infty} \frac{f^{(n)}(0)}{n!} x^n.$$

iii. There exists a power series whose interval of convergence is $[0, +\infty)$.

iv. If Σa_n converges, then the power series $\Sigma a_n x^n$ has radius of convergence at least 1.

v. If the power series $\Sigma a_n x^n$ converges for all $x \in \mathbb{R}$, then the series Σa_n converges.

vi. The convergence interval for the Taylor series of f the same as the convergence interval for the Taylor series of the derivative f'.

3. Consider the function $y = f(x)$ sketched below and suppose that $f(x)$ has Taylor series:

$$f(x) = a_0 + a_1 \cdot (x-4) + a_2 \cdot (x-4)^2 + a_3 \cdot (x-4)^3 + \cdots \quad \text{about } x = 4.$$

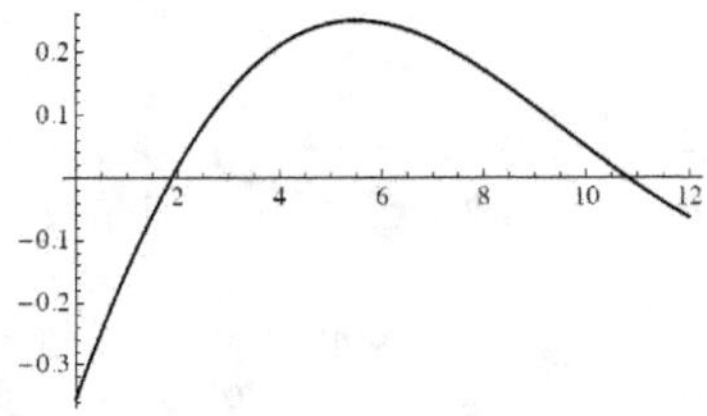

i. Is a_0 positive or negative? Explain.

ii. Is a_1 positive or negative? Explain.

iii. Is a_2 positive or negative? Explain.

4. Find the radius of convergence and the interval of convergence of:

i. $\sum\limits_{n=1}^{\infty} \frac{(-1)^n}{n \cdot 2^n} \cdot (x+2)^n$

ii. $\sum\limits_{n=1}^{\infty} \frac{x^n}{\sqrt{n}}$

iii. $\sum\limits_{n=1}^{\infty} \frac{(n!)^2}{(2n)!} \cdot x^n$

5. i. Find the Maclaurin series for $f(x) = \arctan x$.

ii. Use your answer to obtain a formula for $\pi/4$.

7. Decide whether the following series converge or diverge.

i. $\sum\limits_{n=1}^{\infty} n \cdot e^{-n^2}$

ii. $\sum\limits_{n=1}^{\infty} \frac{(-1)^n}{n^2+1}$

iii. $\sum\limits_{n=1}^{\infty} \frac{1}{n \cdot (n!)^a}$, for $a \in \mathbb{R}$

8. Prove that for all $x \in (-1, +1)$:

$$\int_0^x \frac{1}{\sqrt{1-t^2}}\, dt = x + \sum_{n=1}^{+\infty} \frac{1 \cdot 3 \cdot 5 \ldots (2n-1)}{2^n \cdot n! \cdot (2n+1)}\, x^{2n+1}.$$

9. Using the Maclaurin series for e^x, prove that $|e^{xh} - 1| \leq e^{|h|} - 1$, for all $x \in [0, 1]$ and for all $h \in \mathbb{R}$.

10. Let $\sum_{n=1}^{\infty} (-1)^n A_n x^n$ be a power series such that $A_n = \sum_{k=1}^{n} \frac{1}{k}$.

    i. Prove that the radius of convergence of the power series is 1 and evaluate the interval of convergence.

    ii. If $f(x) = \sum_{n=1}^{\infty} (-1)^n A_n x^n$, for $x \in (-1, 1)$, prove that:

$$(1 + x) \cdot f'(x) + f(x) = \frac{1}{1+x}.$$

Chapter 6

Infinite Products

In this final and short chapter we introduce the concept of infinite products:

$$\prod_{n=1}^{\infty} a_n := a_1 \cdot a_2 \cdot \ldots \cdot a_n \cdot \ldots$$

Our intention is to briefly describe and relate the essence of convergence of infinite products to that of infinite series. Towards that end, we omit the proofs of theorems and we rather emphasize on examples. We encourage the readers though to work the proofs of all results presented in this chapter out by recalling the proofs of Chapter 3.

We recall now some properties of finite products:

$$\begin{aligned} \prod_{n=1}^{k} (a_n \cdot b_n) &= \left(\prod_{n=1}^{k} a_n\right) \cdot \left(\prod_{n=1}^{k} b_n\right) \\ \prod_{n=1}^{k} (m \cdot a_n) &= m^k \cdot \prod_{n=1}^{k} a_n \\ \prod_{n=1}^{k} \left(\frac{a_n}{a_{n-1}}\right) &= \frac{a_k}{a_0}, \quad a_k \neq 0 \; \forall k \in \mathbb{N} \\ \prod_{n=1}^{k} (a) &= a^k \end{aligned}$$

and moreover:

$$\prod_{n=1}^{k} \left(\frac{a_n}{b_n}\right) = \frac{\prod_{n=1}^{k} (a_n)}{\prod_{n=1}^{k} (b_n)}$$

Definition 6.0.1. Let a_n be a sequence. We define a new sequence

$$p_n := a_1 \cdot a_2 \cdot \ldots \cdot a_n, n \in \mathbb{N},$$

the sequence of *partial products*. Then:

$$\lim p_n := \lim \prod_{i=1}^{n} a_i := \prod_{n=1}^{\infty} a_n.$$

Notation 6.0.2. We set $a_n = 1 + u_n$ (for practical reasons that will become clear in the next paragraph; see Theorem 6.2.2) and the infinite product $\prod\limits_{n=1}^{\infty} a_n$ becomes $\prod\limits_{n=1}^{\infty} (1+u_n)$.

6.1 Convergent Infinite Products

Using the analogous arguments with those on infinite series, it would be natural to say that an infinite product converges if and only if the sequence of partial products converges. However, in that case, every product which has one factor equal to zero would converge, regardless of the behavior of the remaining factors. This observation leads to the following definition:

Definition 6.1.1. We say that the infinite product $\prod_{n=1}^{\infty} a_n$ *converges* to $p \in \mathbb{R}\backslash\{0\}$, denoted as $\prod_{n=1}^{\infty} a_n = p$, if the sequence of partial products of a_n converges to p and the infinite product *diverges* if p_n diverges. More precisely:

$$\prod_{n=1}^{\infty} a_n = p \Leftrightarrow \lim (a_1 \cdot a_2 \cdot \ldots \cdot a_n) = p$$

Moreover, we say that the infinite product *diverges* to 0, if $p = 0$.

Example 6.1.2. i. Consider the infinite product $\prod_{n=1}^{\infty} \frac{(n+1)^3}{n^2(n+2)}$. Then:

$$\begin{aligned} p_n &:= \frac{2^3}{1^2 \cdot 3} \cdot \frac{3^3}{2^2 \cdot 4} \cdot \frac{4^3}{3^2 \cdot 5} \cdot \ldots \cdot \frac{n^3}{(n-1)^2(n+1)} \cdot \frac{(n+1)^3}{n^2(n+2)} = \\ &= \frac{2(n+1)^2}{(n+2)} \to \infty \end{aligned}$$

Thus, $\prod_{n=1}^{\infty} \frac{(n+1)^3}{n^2(n+2)} = +\infty$.

ii. The infinite product $\prod_{n=1}^{\infty} n$ diverges, since $p_n = 1 \cdot 2 \cdot \ldots \cdot n \to \infty$.

iii. The infinite product $\prod_{n=1}^{\infty} (-1)^n$ diverges, since $p_n = (-1)^{\frac{n(n+1)}{2}}$.

We now present a result which is straightforward in the case of finite products and is the analogue of Proposition 3.1.3.

Theorem 6.1.3. *The infinite products* $\prod_{n=1}^{\infty} a_{n+k}$ *and* $\prod_{n=k+1}^{\infty} a_n$ *either both converge to the same real number or both diverge.*

The proof is left as an exercise.

Exercise 6.1.4. If $\prod_{n=1}^{\infty} a_n$ converges, then $\exists$ finitely many $n \in \mathbb{N}$ such that $a_n < 0$.

6.2 Tests for Convergence

We now focus on presenting criteria for the converge of infinite products, in a completely analogous way we did to the corresponding criterion for convergence of infinite series.

Theorem 6.2.1 (Cauchy Criteria for Convergence of an Infinite Product). *The infinite product* $\prod_{n=1}^{\infty} a_n$ *converges if-f:*

$$\forall \epsilon > 0,\ \exists\, n_0 \in \mathbb{N} : \left| \prod_{k=n+1}^{m} a_k - 1 \right| < \epsilon,\ \forall m > n \geq n_0$$

Theorem 6.2.1 leads to the following result:

Theorem 6.2.2. *If the infinite product* $\prod_{n=1}^{\infty} a_n$ *converges, then* $a_n \to 1$.

We will use Theorem 6.2.2, the proof of which is a corollary of Theorem 6.2.1 and is left as an exercise, in order to prove that an infinite product diverges just like the Test for Divergence for infinite series (recall Theorem 3.2.6).

Example 6.2.3. Consider $\prod_{n=1}^{\infty} \frac{(n+1)^3}{n^2(n+2)}$. We have that $\frac{(n+1)^2}{n^2(n+2)} \to 0$, and thus the infinite product diverges.

An immediate consequence of the results presented so far is the following theorem:

Theorem 6.2.4. *If* $a_n \geq 1$ *for all* $n \in \mathbb{N}$ *and* $\lim a_n \neq 1$, *then the infinite product* $\prod_{n=1}^{\infty} a_n$ *diverges to* $+\infty$.

Example 6.2.5. Let $a_n = \frac{2n}{n+1}$. We have that $a_n \to 2$ and $a_n \geq 1$ for all n. Thus, $\prod_{n=1}^{\infty} \frac{2n}{n+1} = +\infty$.

We now relate the convergent infinite products to convergent infinite sums:

Theorem 6.2.6 (Test for Convergence I). *If* $a_n > 0$ *for all* n, *then the infinite product* $\prod_{n=1}^{\infty}(1+a_n)$ *converges if-f the infinite series* $\sum_{n=1}^{\infty} a_n$ *converges, i.e.:*

$$\prod_{n=1}^{\infty}(1+a_n) = p \in \mathbb{R} \Leftrightarrow \sum_{n=1}^{\infty} a_n = s \in \mathbb{R}$$

Proof. Let s_n denote the the partial sum and p_n the partial product. Since $a_n > 0$ for all $n > 0$, both s_n and p_n are monotone increasing. Thus, in order to prove the theorem, it suffices to show that s_n is bounded above if-f p_n is bounded above. Obviously, $s_n < p_n$ for all $n > 0$. Moreover,

$$p_n = (1+a_1)(1+a_2)\dots(1+a_n) \leq e^{a_1}e^{a_2}\dots e^{a_n} = e^{s_n},$$

since $1 + a \leq e^a$ for all a. Hence:

$$p_n \text{ bounded above } \Leftrightarrow s_n \text{ bounded above}$$

Finally:

$$p_n \to \infty \Leftrightarrow s_n \to \infty$$

and that concludes the proof. □

Definition 6.2.7. We say that the infinite product $\prod_{n=1}^{\infty}(1+u_n)$ converges absolutely, if the infinite product $\prod_{n=1}^{\infty}(1+|u_n|)$ converges.

Theorem 6.2.8. *If the infinite product $\prod_{n=1}^{\infty}(1+u_n)$ converges absolutely, then it converges.*

The proof follows from the Cauchy criterion together with the inequality:

$$|(1+a_{n+1})\dots(1+a_m)-1| \leq (1+|a_{n+1}|)\dots(1+|a_m|)-1$$

Remark 6.2.9. We have that if $a_n > 0$ for all $n > 0$, then the infinite product $\prod_{n=1}^{\infty}(1+u_n)$ converges if-f the infinite sum $\sum_{n=1}^{\infty} u_n$ converges. Theorem 6.2.8 then can be rephrased as follows:

The infinite product $\prod_{n=1}^{\infty}(1+u_n)$ converges absolutely if-f the infinite sum $\sum_{n=1}^{\infty} u_n$ converges absolutely.

The fact that $a_n > 0$ for all $n > 0$ is extremely important now. Consider for example $a_n = \frac{(-1)^n}{\sqrt{n}}$. We have that $\sum_{n=1}^{\infty} a_n$ converges (*why?*), while $\prod_{n=1}^{\infty} u_n$ diverges (*exercise*).

This leads to the following result:

Theorem 6.2.10 (Test for Convergence II)**.** *If $a_n > 0$ for all n and $a_n \neq 1$, then the infinite product $\prod_{n=1}^{\infty}(1-a_n)$ converges if-f the infinite series $\sum_{n=1}^{\infty} a_n$ converges, i.e.:*

$$\prod_{n=1}^{\infty}(1-a_n) = p \in \mathbb{R} \Leftrightarrow \sum_{n=1}^{\infty} a_n = s \in \mathbb{R}$$

Proof.

$$\sum_{n=1}^{\infty} a_n = s \in \mathbb{R} \Rightarrow \prod_{n=1}^{\infty}(1+|-a_n|) = p \in \mathbb{R} \Rightarrow$$

$$\Rightarrow \prod_{n=1}^{\infty}(1-a_n) = p' \in \mathbb{R}$$

□

Review Exercises

1. Prove that $\prod_{n=2}^{\infty}\left(1-\frac{2}{n(n+1)}\right) = \frac{1}{3}$.

2. Prove that $\prod_{n=2}^{\infty}\left(1-\frac{1}{n^2}\right) = \frac{1}{2}$.

3. Prove that $\prod_{n=1}^{\infty}\cos\left(\frac{x}{2^n}\right) = \frac{\sin x}{x}$.

4. Determine whether the following infinite products converge or diverge:

   i. $\prod_{n=2}^{\infty}\left(1+\frac{1}{n^k}\right)$

   ii. $\prod_{n=2}^{\infty}\left(1-\frac{1}{n^k}\right)$

   iii. $\prod_{n=2}^{\infty}\left(\frac{n^2+1}{n^2-1}\right)$

   iv. $\prod_{n=2}^{\infty}\left(\frac{n^2-1}{n^2+1}\right)$

   v. $\prod_{n=1}^{\infty}\left(1+\frac{x}{n}\right)$

Bibliography

[1] Apostol, Tom., *Calculus 2*, New York: Blaisdell, 1961.

[2] Briggs and Cochran, *Calculus: Early Transcendentals*, International Edition.

[3] Courant and John, *Introduction to Calculus and Analysis, vol.1*, Springer-Verlag 1952.

[4] M. Muresan, *A Concrete Approach to Classical Analysis*, Springer-Verlag, 2000.

[5] I.P. Natanson, *Theory of Functions of a Real Variable*, Frederick Ungar Publishing Co., New York, Vol. I (1955), Vol. II (1960).

[6] S.M. Nikolsky, *A Course of Mathematical Analysis*, Mir, Moscow, 1981.

[7] Rangachary Kannan, Carole King Kruger, *Advanced Analysis on the Real Line*, New York, John Wiley & Sons, 1966.

www.ingramcontent.com/pod-product-compliance
Lightning Source LLC
LaVergne TN
LVHW022110200926
848481LV00003B/1140
9781925823196